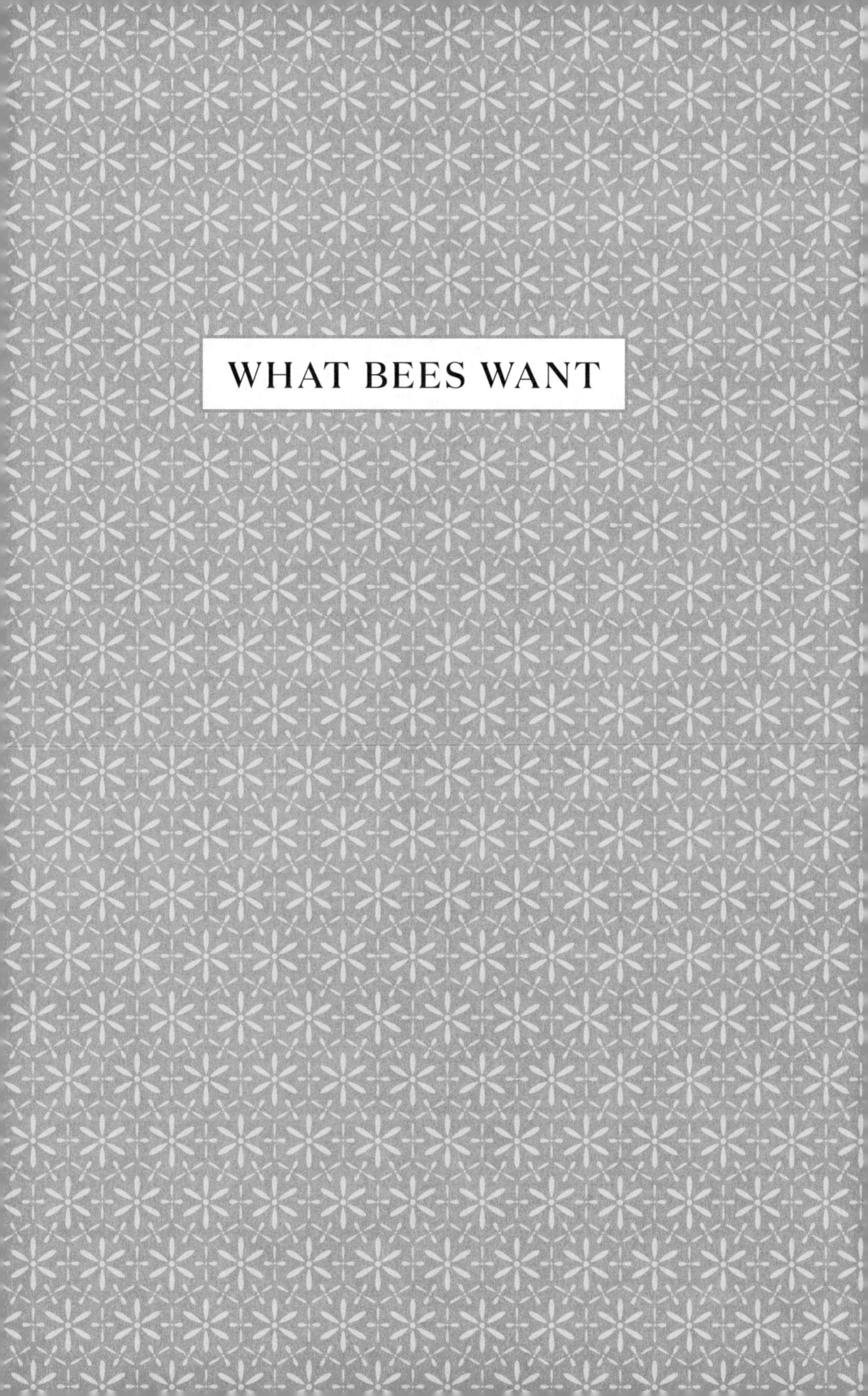

WHAT BEES WANT

WHAT BEES WANT

Beekeeping as Nature Intended

Susan Knilans

Jacqueline Freeman

The Countryman Press

An Imprint of W. W. Norton & Company
Independent Publishers Since 1923

This book is intended as a general information resource about preservation beekeeping. Please read the entire book carefully before undertaking any of the activities it describes. Please keep in mind that bees are venomous insects. To keep yourself safe, you must know how to read their mood. If they are "hot," they are telling you to leave them alone, and you should do so. Using a ladder to catch a swarm is very risky. If you cannot handle a particular swarm without using a ladder, let it go. If you are allergic to bee stings, never go anywhere near bees without an EpiPen® or other epinephrine auto-injector.

Any URLs displayed in this book link or refer to websites that existed as of press time. The publisher is not responsible for, and should not be deemed to endorse or recommend, any website other than its own or any content that it did not create. The authors, also, are not responsible for any third-party material.

Printed in the United States of America

For information about permission to reproduce selections from this book, write to Permissions, The Countryman Press, 500 Fifth Avenue, New York, NY 10110

For information about special discounts for bulk purchases, please contact W. W. Norton Special Sales at specialsales@wwnorton.com or 800-233-4830

Manufacturing by Versa Press
Book design by Chrissy Kurpeski
Production manager: Devon Zahn

Library of Congress Cataloging-in-Publication Data

Names: Knilans, Susan, author. | Freeman, Jacqueline, author.
Title: What bees want : beekeeping as nature intended / Susan Knilans, Jacqueline Freeman
Other titles: Beekeeping as nature intended
Description: New York, NY : The Countryman Press, [2022] | Includes index.
Identifiers: LCCN 2021049590 | ISBN 9781682686737 (cloth) | ISBN 9781682686744 (epub)
Subjects: LCSH: Bee culture. | Handbooks and manuals.
Classification: LCC SF523 .K58 2022 | DDC 638/.1—dc23/eng/20211028
LC record available at https://lccn.loc.gov/2021049590

The Countryman Press
www.countrymanpress.com

A division of W. W. Norton & Company, Inc.
500 Fifth Avenue, New York, NY 10110
www.wwnorton.com

10 9 8 7 6 5 4 3 2 1

"Bees are of an entirely different order.
Their journey is not the same as ours."

—MICHAEL JOSHIN THIELE

"It is apparent to me that modern beekeeping will eventually tear the species of honeybees into its demise, if no other counter movement will arise. Not because beekeepers are vicious or ill minded but because the majority does not understand the effects of their manipulations as a whole."

—TORBEN SCHIFFER

"The word 'dominion' in Genesis is often interpreted as divine sanction for all forms of animal abuse. But the original meaning of the word 'dominion' comes from the Hebrew verb *yorade*, which means to come down to, to have communion with and compassion for."

—DR. MICHAEL W. FOX

This book is for my husband, who has helped me gather swarms and carry hives, and who has come to love the bees as deeply as I.

A special dedication also to Torben Schiffer and Tom Seeley, for generously sharing their research with us. You bring the science to our methods, and give voice to the wildness of bees.

—SUSAN KNILANS

For my dear mother, Jesse Carole Stone, who, from the beginning, encouraged me to explore and love the land, identify the floral bounty and native wildlife, gather and cook field-foraged weeds, climb a hundred trees, draw and paint my visions, and to have the courage to be who I have become. It really was okay with her that we once had 32 cats.

—JACQUELINE FREEMAN

Contents

Foreword: The Road to Freedom

Torben Schiffer, Dipl. Biol.

I have been keeping bees since 2004. In my early years, I learned the concepts and practices of conventional beekeeping. But each successive year, I learn more about a different path, one that shows me the bees' delicate relationship with nature. To free my bees, I first had to free myself from conventional beekeeping.

A few years ago, my bee education and appreciation took a big leap as I began to understand how in many ways humans have thwarted natural bee behavior. I was taking a walk down the tree-lined dirt path from my house to the organic forest plantation a few minutes away. The sun shone through uncountable gaps between the branches and leaves of the old apple trees. Deep in thought, I suddenly heard a swarm. Thousands of bees flew high in the sky, a dense cloud of buzzing honeybees about 20 meters above my head.

The bees headed across my neighbor's yard, heading in the direction of my own home. I ran across my neighbor's property, chasing the bees. When the swarm reached the giant spruce tree in front of my house, the bees took a sudden right and headed directly to my backyard. I ran toward my house and—for the very first time—

witnessed approximately 20,000 bees swarming to the balcony above my porch.

The summer prior to this, I had placed three simulated tree cavities (which I call SchifferTrees) on the balcony outside my research office, where I could easily watch the bees come and go. Two of the SchifferTrees had remained empty as I waited for local bees to find them. Now, new life was about to move in.

I ran into the house, grabbed my camera and flew up the stairs as thousands of bees arrived at their chosen hive and soared everywhere around it. I stepped outside the balcony door and was enveloped by the buzzing cloud of thousands of bees. I became a part of them. Every fiber of my being connected to an indescribable force of nature. I felt the bees' collective agreement and single-minded determination to move into their new home.

My neurons felt charged with the electricity of the moment. The freedom of these bees had been a long time coming. I had intentionally purchased the colony from a conventional beekeeper as part of a rewilding project I was working on, and this swarm had come out of one of the boxes I placed on the forest plantation a few weeks earlier.

These colonies had been kept under the rules of modern conventional beekeeping, which means they were never allowed to express their natural behavior. With their previous owner, the bees lived in flimsy wooden hives and served solely as livestock who produced honey. Now, never again would they have to live in these unnatural conditions. Witnessing all the colony's evolutionary abilities and natural behaviors playing out for the first time was electrifying and filled my heart with happiness. I experienced such a sense of relief and contentment. This was, for me, a truly overwhelming and spiritual event.

After those bees had settled in, I hurried down to the plantation again in order to check on the other hives. I almost couldn't believe my eyes—the next swarm was beginning the minute I arrived. I started recording, and the footage shows me enthusiastically talking to the bees, guiding them, warning: "If you move into the other tree, I will lose you!"

Then, thousands of them started flowing out of the entrance gap. They were like a constant energetic stream of water that would instantly evaporate into the sunlight. Another huge cloud of countless buzzing, freedom-seeking honeybees was inviting me again to witness their truly breathtaking abilities, honed over millions of years.

But with this colony, something was wrong. The cloud of bees didn't leave the forest plantation and they didn't continue soaring over to my house. Suddenly the whole swarm returned to the box, turning it brown with the masses of their bodies. Their buzzing was slowly fading, and everything about them conveyed uneasiness.

I spotted a handful of them lying in the grass right in front of the hive. And there she was: the queen, lying in the dust unable to fly, eagerly yet vainly trying to get airborne and follow her daughters to their next home. This was another consequence of the unnatural practices of conventional beekeeping playing out before my eyes. Production beekeepers will often cut the queen's wings, a tactic meant to prevent the hive from swarming, as a colony will never leave their queen. Crippled by this conventional practice, this beautiful creature would never fly again—and nor would her colony, dedicated as they were to her wellbeing.

I picked the grounded queen up and enclosed her in my hands. There she was, trying to escape through tiny gaps of my fingers. One time she made it out, tried to get airborne, and fell down into the grass again. My euphoria crumpled into pure sadness, deeply sorry at what my species had done to her. Luckily, seven days later, her virgin queen daughters hatched and formed three new swarms. Their colonies developed well and are part of my rewilding project now.

There is a growing international movement of people wanting to care for bees as nature intended, allowing them their full integrity of being, and honoring their right to their own evolution. When I first met Jacqueline and Susan at the Learning from the Bees Conference in the Netherlands, I saw instantly that they, too, were forging ahead on a path of knowledge that led to helping bees live as nature intended.

Jacqueline and her husband Joseph came to stay awhile at my home near the Black Forest in Germany. I had just received a high-powered video camera that was used for micro-surgery, and we immediately put it to use watching the bees doing their daily tasks and how they lived inside my tree-cavity SchifferHives. On the screen, each bee was as large as my hands, every detail clear and crisp. We marveled at their orderly communications with each other, with other insect inhabitants in the hive, and how they all worked together in peace and harmony.

Susan grabbed me up during a lunch break at the conference and showed me photos of her skeps (woven hives) and log hives. Weaving skeps is an undertaking of great love and great patience. Keeping bees in the way of preservation beekeeping requires a lot of study and commitment. I immediately recognized in Susan a kindred spirit, who, like me, was willing to go the distance and buck the norms of convention to help honeybees, who are struggling so much across the globe.

Since the conference, I have been so humbled that these two thoughtful and caring bee lovers are using my research to guide their work with bees, completely recrafting their beekeeping classes to reflect what bees need.

It is not easy to let the bees guide our knowledge, but that is truly what we need to do to understand what is important and necessary to them. This book is an important contribution to the world of bees and humans. Bees are so much more than the honey they produce. The pages that follow reveal what is normally hidden or unnoticed by most beekeepers. To see these hidden wonders of the hive, we must learn to see with the heart's eyes. Only a true love for nature and the honeybee can coax out those secrets.

Preface: Following the Bees

She lands quietly on the back of my hand where I kneel pulling grass out of the garden beds. I look down at her, pause, and set aside my weeding tool. Surely, gardeners should stop and smell the roses now and then. Bringing my hand up near my face, I watch this small foraging honeybee stop to get herself in order before taking the 10-foot flight to her hive's landing board.

For a moment, it seems all her six legs are moving at once, raking pollen dust down to her hairy thighs where she packs it in hard lumps with just a drop of fresh nectar. I watch her abdomen pump like a tiny accordion, settling her breath after a morning of ambitious flower tasting. Now, her amber tongue comes out like a finger, touching my hand. I see the little shiny tube-like tongue lick across my skin. Her antennae wave curiously. I've always imagined that this is how bees talk to themselves when they are alone, the antennae a running monologue, never missing a beat.

Certainly, she knows me. Honeybees can recognize faces, and they easily note strangers to their environment. She is as foreign-looking as any insect, and yet I've come to feel such mammalian affection for this fuzzy little being and all her thousands of sisters. I extend a finger, and she meets it with her own tiny foot, then helicopters up and away. I pick up my weeding tool with a smile, and tackle the dirt refreshed from just this small chance encounter with grace.

In your hands is a celebration of bees. I began keeping bees years ago, and I would dread any spring that didn't include the sounds of bees in my yard. I'm not into bees because of honey. I'm into bees because they have captivated me. They do amazing things with my yard, making a small Eden out of my talentless gardening efforts.

The sound of them, the scent of them, the sight of them dipping and gliding like leaves on the wind enchants me and heals me in ways that defy words.

My bee journey began literally at the feet of my co-author, Jacqueline Freeman, author of *Song of Increase*. I sat on a floor cushion at her farm in rural Washington, enrapt, as she shared with us the bee wisdom she had acquired through many years of respectful observation. Unlike most beekeepers at the time, Jacqueline was modeling an entirely different way of tending bees. It was a practice in which the bees' needs came before the keeper's, as well as before the keepers' desire for honey.

Our meeting was a match made in bee heaven: Jacqueline was a gifted, passionate teacher. I was an enthusiastic and serious student. Soon, I was teaching the beginner beekeeping series on my own. I helped to compile and edit Jacqueline's first book, *Song of Increase*, soaking up the wisdom of the bees along the way. Then, Jacqueline and I realized that after the beginner classes, our students had nowhere to turn to continue their path in natural beekeeping.

So, we partnered together for a Bee Club. Less than a year later, we co-founded the nonprofit "Preservation Beekeeping Council" and grew our community. Though our nonprofit ultimately closed its doors during the 2020 COVID-19 pandemic, we still work together on local bee-enhancing projects like making and installing natural bee habitats on private and public lands, offering classes, planting sidewalk strips with bee forage, and rescuing and rehoming swarms.

In 2018, our Preservation Beekeeping Council led us to Holland where we attended the first International Natural Beekeeping Conference, "Learning from the Bees." There were hundreds of beekeepers from 32 countries, every row speaking a different language. We wore headphones as translators deciphered foreign words for us. Amid all those languages, we shared the vision of making the world better for bees on their terms. Jacqueline was a speaker. I spent the breaks teaching folks how to weave old-fashioned skeps (spiral-shaped straw hives). Holland was a game changer for us, revealing new science that fully supported the ways we had always tended our bees.

A CASCADE OF CHALLENGES FOR BEES

You would have had to be living under a rock for the past decade to not know that bees are struggling worldwide. Agribusiness and Big Pharma would like you to think that the largest threat to bees is the Varroa mite. This stowaway insect from the Asian continent did kill multimillions of bees in the United States when it first made landfall in the early 2000s. But the Varroa mite is not the only culprit.

In reality, pesticides and poisons actually lead the charge in the threat to bees. And sadly, the next greatest danger to all to our honeybees are those beekeepers who use conventional production management practices.

Yes, you read that correctly. Conventional beekeeping practices focused on scale and production pose a great risk to bees. These may seem to be fighting words, but let us explain.

Beekeepers today can be divided into one of two camps: production beekeepers and other beekeepers. Those in the former camp manage bees as a part of the commercial system and, as a result, they have completely different motivations and yearly goals from most backyard bee tenders. Their goal is largely to profit from products of the hive.

These production keepers follow a now worldwide system of management that has developed over the past 150 years. The practices used in this system are designed for maximum honey production. In the US, production keepers are subject to all sorts of regulations and oversight by the US Department of Agriculture, as they should be. Their practices include chemical treatments in the hives, frequent hive manipulation techniques, and aggressive breeding programs that separate the young queens from their natural cohort, turning them into commodities rather than families.

Only recently are we learning what focusing on honey and money have cost our bees. Much of what is done in conventional beekeeping on a weekly basis is actually harmful or even lethal to bees. In fact, production keepers expect to have great losses each year. And the numbers are staggering, with over 40 percent bee loss in 2019, the highest ever recorded.

All the bee schools and state beekeepers associations currently in place are a product of the USDA. So for beekeepers who want to profit from products of the hive, their rules are clear. But for all the other beekeepers, there are almost no programs, or clubs, or support systems to show them the ropes. While many beekeepers want to do right by their bees, there is very little instruction available.

But an exciting new way of being with bees is emerging, championed by bee tenders like Jacqueline who were listening to the bees, observing them in wild hives, and questioning the USDA's bee policies. What Jacqueline and I call Preservation Beekeeping is one of the many new forms of what is loosely referred to as "natural beekeeping."

Preservation Beekeeping asks simply: What do bees want? How can we help them achieve their maximum life force? With this guiding principle, we determine how to best create the most bee-appropriate hives and bee gardens, with as little interference in the bees' lives as possible.

Our way of caring for bees is far less stressful for them (and for their tenders). Please don't assume our hands-off management implies we don't know our bees. We regularly observe our bees, read current bee books, participate in near-daily discussions with other bee colleagues, stay abreast of current research, and draw conclusions fueled by consistent continued education. We steadily innovate our hive styles and work to maximize forage in our bee gardens. We know each colony is an individual, with a unique personality and preferences, and we treat them accordingly.

A DUAL-PURPOSE BOOK

This book is both a memoir of my own bee journey with Jacqueline and a handbook for our style of Preservation Beekeeping. We initially wrote this book as a simple "how-to guide." But our method of bee tending is deeply relational, and we wanted curious readers not to just learn the nuts and bolts of our management style, but to fall in love with bees along the way. I hope the tales of our hives and the colonies that populate them will find their way into your heart and body, and bring you as much healing and delight as the bees have brought to Jacqueline and me.

In the pages to come, you will learn about the best hives for bees, where to place them, how to learn from your bees, and how to plant for and tend your bees. What you won't find here is a step-by-step methodology to follow week-to-week or month-to-month.

Rather, we offer you the cornerstones of bee-appropriate, preservation-style, hobby bee tending, and show you ways to apply these tenets in your own yard with your own bees. There is much room for creativity here. Bees are stunningly complex, varying their activities and preferences with their particular geography and weather patterns. There is no such thing as "one size fits all" when it comes to bee tending.

We have many, many good examples to set you off in the right direction, and over time you will find that the bees themselves make it clear what works for them and what does not in the particular landscape of your own yard.

In the process of caring for bees, we care for ourselves and our planet. What bees want is safe housing; clean, untainted food and water and soil; and to live their lives as nature has crafted them to do. In giving our bees what they want, we build a vibrant, more abundant world.

We ask nothing from our bees, but take our joy in supporting them and hearing them sing their Song of Increase. Today, many people imagine they would like to benefit bees in this way. For them, for you, we've created this book. As you follow the principles in this volume, you will become a fine friend to bees.

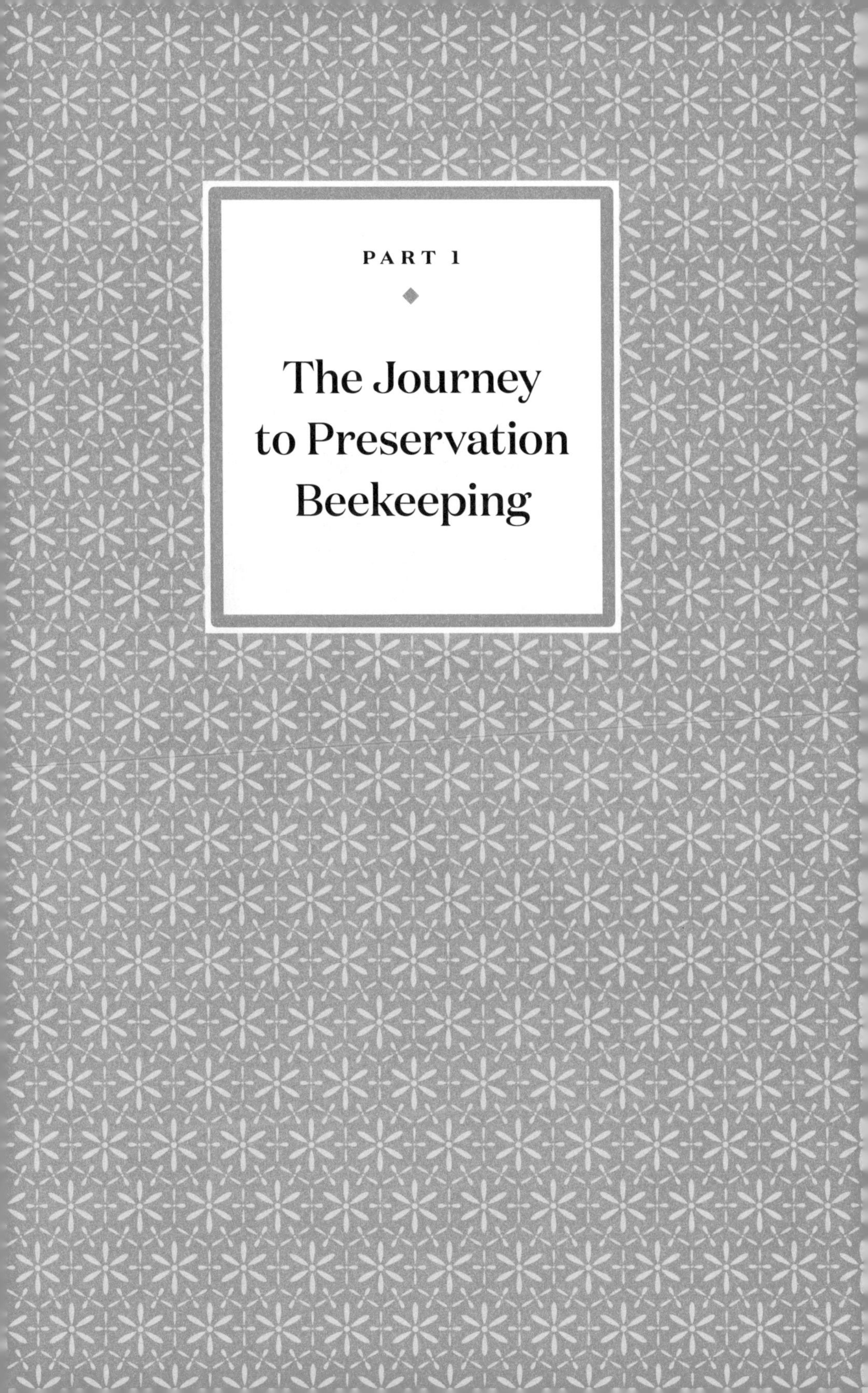

PART 1

The Journey to Preservation Beekeeping

1

When the Bees Came

I climbed over the edge of tall scaffolding, 20 feet in the air, and found 10,000 bees glittering in the morning sun. The historic church I stood next to had been a blessed home to this treasure for nearly 50 years. The bees fluttered their wings in unison and their sweet scent wafted around me.

Renovations had begun on the old church, where bees had dwelled in the eaves for longer than the congregation could remember, and the hope was that they could stay in place. My bee teacher Jacqueline Freeman had been out earlier to see if there was any way the workmen could work around them, but the workmen pointed to water damage that needed fixing in the roof near their hive. It was clear the bees would have to be removed. So Jacqueline put out the call for more hands. Beekeepers Bill and Tel volunteered to run the removal and I quickly offered my help.

Mine were, however, the most unskilled of hands. That summer, I had—on a whim—visited Jacqueline's farm to take a class on bees and beekeeping. A year earlier I had moved from Indiana back to the Pacific Northwest, having spent the previous years working at a wildlife rehabilitation center. But my health and energy levels were simply not up to the tasks of midnight feedings, difficult medicating, and the grief of losing so many struggling animals.

I was actively seeking a way to express my love for the living world. That longing guided me to Jacqueline's farm and to bees.

Jacqueline had been teaching natural beekeeping classes for over a decade at the time, one of only a few beekeepers in the country doing so. Her exuberant and contagious enthusiasm made me want to surround myself with honeybees. Like her, I wanted to see the world from the bees' perspective. I eagerly leaped in. I spent hours reading every bee book she recommended and I took every class she offered. Twice.

I was brand new to bees. Previous to meeting the church-roof hive, my only contact with bees had been one single sting as a child. My mother said the simple swelling meant I was allergic to bees and told me I could die if I was ever stung again. I didn't go barefoot for the rest of my childhood.

When I saw a poster for a "Spirit of the Bee" workshop at Jacqueline's farm, I attended out of curiosity more than anything else. Bees and spirituality? The title proved to be more prophetic than I realized. By the end of the workshop, I was a convert with a near-religious zeal for fuzzy little honeybees.

I've since learned that many people are fascinated by stories of the true nature of bees, and that suits me because I have a lot of amazing stories about honeybees to tell. In these past years since the bees came to me, I've reflected often on why they are so compelling and engaging. When I share my joy over spiders and soldier flies, people often cringe or shake their heads: "Nope, nope, nope!" Yet when I say the word "bees" at the grocery store, the bank, the yoga studio, or the gas station, everyone lights up. "Bees!" they smile broadly. "Oh, we *love* bees!" Positive exposure in the media has helped drive that public enthusiasm, but bees themselves can turn that initial curiosity into a hobby filled with great purpose.

But on that morning with my toes gripping the ladder, I knew none of this. My bee journey was at its beginning. "Come up and take a look," said Tel. "You'll be up close with all of these bees, so you might as well introduce yourself." Tel was tall, soft-spoken, and reassuringly experienced in bee rehoming. I climbed onto the scaffold and edged up to the exposed eave.

Immediately I heard the sweet hum of diligent working, bees. They took no notice of us at all and continued in their labor. The sun beamed and my heart filled. But bee gear is clumsy. I wore a bee-proof jacket with a large hat and veil. It was hard to see the bees clearly through the black mesh—it was rather like looking through Jell-O.

I moved closer to the eave, in awe of the colony's sound, its size, and its otherworldly presence. When examined in such intimate closeness, the marvelous features of insects like this can still be alarming. They are small, scaled, winged. Most people see these surface characteristics and struggle to relate to insects, to see the aspects of life we share: our common needs for water, shelter, food, reproduction. Unlike relating to puppies and bunnies, it takes more mental work to find common ground with creatures sporting scales and stingers.

Bees are more startling in their otherness than many insects. Honeybees are of the rare order of being we call a superorganism. This means they live two lives: one as a singular bee, the other as a tiny component—a cell, you might say—of a much larger system. Peering at the bees through the mist of the veil, I was meeting them in their entirety, and it was sobering. The colony was huge, old, and thriving. They seemed happy with their home. I wondered how they would feel about being forcibly removed from it.

Although a healthy dose of caution would have been appropriate, I was not scared. I was tingling, the excitement running through my body like an electrical current. I had never seen bees like this, shimmering in a great living wall, their hum vibrating through my body as I stood only 2 feet from them. Since childhood, I'd had no further contact with bees, and certainly no experience with bees by the tens of thousands.

"This is what we are going to do," Tel instructed. "Bill and I are going to start at one end of the eave and work to the other." He described how he and Bill would cut out combs from the hive. They would take the ones full of honey and place them in one bucket, and place combs full of brood (unborn bees) in the other. They would then lower the buckets down to me, where my job was to put all the honeycombs into a large plastic tub and, with a string, tie the brood combs one by one into the frames of the hive. "You don't need to be graceful in the tying. You just need to remain calm and focused. And be sure to tie the combs right side up, that keeps the brood and honey settled in."

Where would the bees live? We had brought several empty hives for the bees, including a Warré. A Warré hive is designed to be more like a vertical hollowed out tree than the hives conventional beekeepers use that are designed for maximum production of honey. But we didn't care about the honey harvest. We wanted to replicate a natural bee home as much as possible, so a Warré was a good choice for the church colony. Inside the Warré, wooden strips, called "bars," lie across the top. And it is on these bars that the bees affix their long hanging wax combs. These bees weren't building from scratch, however. My job was to do the best I could to put these combs together in an orderly fashion, as they would appear in a natural hive. The goal was that I would put this beehive back together again in some semblance of a natural home that met all the bees' needs.

Once I had filled one box, I was to stack another fresh box on top, and keep building, like stacking sections of a log, one atop the other.

Feeling overwhelmed, but eager for our plan, I crept down the ladder and hurried to prepare for the bees. Bill set up a card table as my workstation. On it were spools of string, scissors, bowls of water to rinse off my hands when they got too sticky, and paper towels. I had also brought along a box of thin latex gloves that I thought might prevent bee stings.

Having remained sting-free all of my adult life, I had no idea what a bee sting would do to me. In retrospect, it was supremely foolish of me to approach a day of working with bees while not knowing if a sting might put me in the hospital. Impulsiveness has always been a trait of mine, and I count myself lucky that I'd not felt too many awful consequences on my life's many hasty decisions.

I pulled on the thin gloves and began cutting lengths of string that would tie the combs to the narrow wooden strips. Overhead, Tel and Bill began working with long knives cutting out full slices of comb. A white 5-gallon bucket appeared over the edge of the scaffold board and they lowered it down to me on a long rope. I held up my arms and embraced a full bucket of bees and comb.

I set the bucket on the card table and peered inside, shivering with excitement. I hesitated, unsure of where and how to begin with my portion of the plan. Bees lifted gently into the air and formed a sheer curtain of amber around the table. The bucket vibrated with a deep thrumming hum-a sound that would fill the rest of the day. They covered the combs completely, their bodies continuously rippling and moving like water flowing.

Gingerly, I started to lift a slice of comb from the bucket. At that moment, I did not expect warmth. While I knew from my classes that bees have the capacity to regulate the temperature inside the hive to a balmy 90° to 98°F, this fact hadn't fully registered. As I picked up the first comb, my hands filled with heat. In my palms, the bees felt like fuzzy warm kittens.

There was still so much at this point, so young in my beekeeping career, that I didn't know. I had no knowledge, then, that the wax combs are naturally soft and flexible because of the warmth of the bees. I also did not know that bees will cover an entire comb of living brood and do their best to stay on top of it. Instead of leaving, each bee spreads her body over a bee larva to maintain the baby bee's heat and protect every cell. Tel had told me I'd be the one tying the brood comb onto the hive bars. He didn't mention that the bees would remain clinging tightly to each side of the comb.

I had to get moving. More buckets were ready to lower from the scaffold. As I worked to attach each soft wax sheet, the comb drooped over the edge of my palm, like one of Salvador Dali's watches. If I pressed too hard with my fingers on the comb's underside, the bee under my finger would sing out a sharp "Nyeeet!" Instinctively, I began singing, reassuring myself and the bees that I was up to the task and that we could work together. The constancy of the singing settled the bees, settled me.

I placed two pieces of string parallel onto the surface of the card table. I set a wooden bar perpendicular across it. Then I laid the wax comb, shimmering with bees, onto the string, and began tying each long comb onto the bar. In the roof cavity of their original hive, the bees had glued the tops and sides of the combs so they were secure, but in removing the combs, the bars served to give them a sturdy structure again. It was tedious work and immediately I realized my gloves had to come off. They made my fingers too clumsy for tying and for gently coaxing bee bodies away from the strands. I struggled to remember the verbal directions now that I had a thousand bees in front of me. What was that about there being a right-side-up on each comb? Oh, dear. Softly I blew on a small section of the comb, thinking I was gently asking them to move over so I could see what end had been cut away and figure out which cut end should be the "up" end.

Here's another lesson I learned: Honeybees don't care for the scent of human breath. Bees have an incredible sense of smell (see page 102) and CO_2 gives them the heebie-jeebies. They will communicate their distress to you by flying up and stinging you if you puff at them too much. These bees were kind to me, however, and I learned to give a very brief puff to the edge of each comb so that I aligned it right. I was looking to see which way the comb's cells were oriented to keep liquids and baby bees securely in the cell. If I mistakenly tied the comb upside down, the baby bees would be upside down, as well.

What else did I learn? That bees tickle! Across my palms and the backs of my fingers, bees wandered with curiosity, their fuzzy bodies and small feet brushing against the fine hairs of my hands. The sensation was oddly comforting: warm, furry, and scritchy. Some bees licked my fingers (sweat? or was that a lick of honey?). The sounds coming from the bucket were a happy kind of white noise and I fell into a lovely, contented trance.

In years since, I have learned that people can pay to sit next to beehives and breathe the air from the hives with a special mask. Studies have shown that the sound and scent of bees is relaxing to the human nervous system, and can help soothe many of the symptoms of PTSD.

Bent low over the card table with the sound and scent of the bees wafting all around me, I knotted strings onto wooden bars for hours, placing each comb in the hive box, and moving bees gently with my fingertips while singing.

My fool's luck held. I had nearly no reaction to the four stings I got that day. Only four stings from 80,000 bees! Mom was wrong: I had never been allergic to bees.

As we closed down our project for the night, I was uncharacteristically silent. The little I had known about bees had been turned on its head. In the midst of having their colony completely removed from its foundation, the bees treated us gently, focusing all their efforts on caring for their young who rested in the sealed brood combs. Their song remained sweet and encouraging, understanding somehow that we were doing our best to help them survive this move to a safer place.

I have had many, many creatures pass through my hands in my life. As an animal zealot, I've soothed and tended lion cubs, bears, elephant babies, and all the local fauna found in cities. Whether mammals or birds, those animals all seemed to exhibit a certain similarity to myself—similar feet and similar faces—and felt familiar to me in a very basic way. Not so with the bees.

Even then, when my knowledge of honeybees could not fill a drinking straw, my immersion in them that day introduced me to an intelligence far different and far deeper than mine. Ancient, ordered, and imposing in their utter mystery, the bees reached far beyond my senses, hooking me with their complexity and wonder. I ended the day realizing I did not know who bees were, and I wanted to remedy that. Such a complex creature would be a profound teacher, perhaps showing me the world through insect eyes.

Though I had dabbled at the edges of bee-ness with classes and books, on that day I turned a corner. I longed to be closer to them, to learn more about them. I wanted a hive in my own yard.

2

The Value of My Mistakes

In your early days of beekeeping, you may find yourself baffled by all there is to learn. If my own experience is any proof, you'll likely make a few mistakes, but so long as you have good intentions, you'll learn and continue to improve your style of beekeeping for the sake of your hive.

I didn't have to wait long for my first hive. The removal of the church bees took us days, and in the end, we filled three hives with bees. As we loaded the last of our supplies at the end of the day, Bill approached me. "Would you like one of these hives in your yard? I need to find a home for it and my yard is near-full. You could care for them, and I can answer questions for you. What do you think?"

My prayers had never been answered that quickly. By nightfall, a colony of the church bees found their home in my yard and I was ecstatic. I named the colony Melissa, and began a journal to document our life together.

The Melissa colony arrived at our home, MillHaven, before we had moved there ourselves. We were temporarily living in a rental while we remodeled the house, every day making MillHaven more livable. I felt such joy seeing the bees busy in the yard on the flowers, working to create a new home for themselves, just as we were.

It is one thing to read about bees, but it is something else entirely to be in the presence of bees. Every bit of practical knowledge I knew about bees simply flew out the window when I sat next to the hive. Like many new beekeepers, once next to a thriving colony, I found their extraordinary presence all-encompassing, leaving my head with little room for facts or thought.

Jacqueline had told me that it took her a thousand hours of observation to feel like she understood even a little, so I started in. She cautioned that observation hours shouldn't be spent constantly identifying what I thought I saw, but open-minded hours spent simply watching. For myself, I was fascinated by every tiny detail: See the bee walking! See the pollen on her leg! Look at that bee fur! See those two bees touching antennae! Every so often the taskmaster part of me bubbled to the surface and I asked myself if there was something I ought to be doing. Is there enough food? Should I feed them? Is the hive in the right place? What are they doing?

I spent hours sitting by the hive on a small stool, watching the Melissa bees and talking to them. For the most part, they took no note of me, coming and going from the hive in a quick and purposeful fashion. I brought out a magnifying glass and watched them close up, noting how finely segmented their antennae were, and how they spoke to each other by touching and stroking each other. In conversation their antennae worked overtime, bending, reaching, touching, and brushing.

How would it feel, I wondered, to have these lovely slim fingers sprouting from above my nose like an extra set of receptors to sense the world? What does the world look like through a bee's eyes, each one containing thousands of lenses that see colors beyond our own perception? I couldn't even imagine.

I didn't yet know bee behavior by the seasons and I made some grave mistakes. Bill sent me home with a bucket of honeycomb to feed the bees while they worked away at making new wax combs and repairing the ones I'd hung in the hive. Filled with good intentions, I placed a huge bowl of honeycomb right on the landing board of the hive so they would not have to travel far for a meal.

This is something that would have been harmless in the cool days of early spring when there are few bees in the air. But it was summer now, and hot. I walked out to check on the bees and was alarmed to see bee wars on the landing board. I called up Jacqueline. "There is all this fighting going on outside my Melissa hive!"

"Susan, take the honey bowl away," she said, reminding me of the bees' cyclical lifestyle, "Remember 'bees by the seasons'?" Jacqueline explained to me that you can only put out food in the very early spring, or you will attract every bee in the countryside to your hive, and they will do their best to get in and rob all the nectar out of the hive. In essence, that bowl of honey announced to every bee in three miles that there was a hive to plunder. My bees were too new to have a strong population of hive defenders yet, and all the other bees in the area knew that the Melissa hive had limited defenses.

Feeding my bees is a practice I have greatly improved upon and one which we detail on page 147. The trial and error of beekeeping, though heartbreaking at times, is always teaching us new solutions. I removed the honey bowl, and with Jacqueline's help, I managed to find a way to place a smaller bowl inside of the hive where it would not attract attention.

Late that afternoon when the sun was still hot and the shadows long, I sat on my stool, trying to broadcast my apologies to the Melissa bees. Mea culpa, mea culpa. Suddenly, a bee flew up from the landing board and gently tapped my head. I smiled, imagining that she was sending me a message that she heard me and all was forgiven. Then another bee tapped the side of my face. Lucky me—two forgivenesses! I raised my hand to my cheek in acknowledgment as another bee flew up and drifted back and forth right in front of my face.

They are communicating with me! We're having a special moment! My reverie was cut short as four bees, buzzing loud and sharp, sprung at my face and stung me. I bolted and ran for the house as bee venom fired up my cheeks.

Pulling the stingers out of my face, I had a painfully sheepish moment remembering Jacqueline describing how bees often try to warn away human misbehavior (see page 126). "Bees will let you know to keep away with several, gentle warnings," she said. "First by flying close with a high-pitched buzz. If you don't back up immediately, another bee will bump you with her forehead, demanding that you back up. If you ignore the messages and are still there, out come the stingers. It's their way of setting a clear boundary and getting you to step back or go away,"

Without me understanding, the bees had provided me with all of those early warnings, starting with speedy fly-bys, flitting back and forth in front of my face, then a quick forehead tap or two. Yes, the Melissa bees had indeed been communicating with me. They were not in the mood to visit and were telling me to leave. Over the next days and weeks, Melissa communicated to me in this bee language that I could sit with her as long as I wanted in the morning hours, but in the late afternoons when the sun was brutal and the colony activity ramped up to full power, they wanted me to get out of their space and stand far away.

Later that summer I went over to Jacqueline's farm to spend the day working with her bees. For her, "working with bees" mostly means watching them on the landing board, at the bee watering station, or in the flower fields where they forage.

Jacqueline and I spent some time watching a first-year top bar hive that had been on the farm for a few weeks. We wondered if they had built sufficient comb and were taking part in the flush of early summer flower nectar and pollen. The morning was cool, peaceful, and calm, and we thought about going in for a look.

Jacqueline has very gentle bees that she has worked with for many years. They know her and are easy about her presence, enough so that she'd stopped wearing a protective bee suit many years ago. We watched at the entrance and decided this would be a good day for a peek at the new hive. And because they were less familiar, Jacqueline handed me a hat with a veil before donning one herself.

Jacqueline slowly removed the top of the hive, exposing the bees below.

Suddenly, bees poured out of the hive, diving at our hats, the veils, and our sleeveless arms. Jacqueline flung a towel to cover the opening. We dashed 40 feet away before we turned to look back. Jacqueline's eyes were like saucers. "I've never seen them like that! Why are they so cranky?"

It was as good an example as any that sometimes we forget how to think of the bees' environment. The cool morning, correct. The sunny sky, perfect. The quiet landing board, just right. So, what went wrong?

"The 4th?" I said.

It was the day after Independence Day. The night before we'd sat on the porch of the farm and watched fireworks exploding around us. Despite being far out in the country, her neighbors put on a wingdinger of a show nearby, scant hundreds of feet from the hives. Every sonic boom—and there were many—vibrated through the air and shook the ground.

"Oh, of course!" she replied. "How could we have forgotten! My poor bees must have thought we were coming into the hive with tanks and cannons." Covering her arms (17 stings!) she went back to close up the agitated hive. Clearly, they were in no mood for an open house that day.

In my yard, as spring turned to summer, the sun showed me I had placed Melissa in the hottest part of the yard. In the Pacific Northwest, the western suns of summer are witheringly hot. Bees work constantly to try to maintain a consistent hive temperature to keep the babies intact. Too hot can be nearly as bad as too cold. In a heat wave, the colony may sometimes send a large bee contingent outside the hive to beat their wings at the entrance and make a convection current to draw out the heat. On hot summer evenings, it looked like half the Melissa colony was hanging out on the face of the hive. I imagined the interior of the hive must be a furnace.

I told Bill I had a better place for the hive. I thought that moving it would be a major operation of some kind, but Bill showed up in his bee suit, and carried the hive box and bees about 50 feet across the yard, placing them on a stand of cinderblocks I'd set up. For the next three days, the Melissa bees were in utter confusion, returning from foraging flights to find their hive simply gone. I thought the scent of the hive would lure them home, but they had been moved just far enough that the hive was utterly off their radar.

I stood in the yard, trying to waft bees in the right direction of home, flapping my arms and fanning with an old magazine. In the evening, I gathered up the little clumps of Melissa bees taking refuge on leaves and flower stalks, and carried them to the new hive site. Each night, more foragers remembered and fewer were at the old site. Then a few nights later, a hard summer rain beat down every tall plant in the yard. Whatever bees had not made it home by then perished.

This was a devastating way to learn how not to move a hive, and I wished I had done it better. We hope we can provide you with clarity on this subject, to guide you away from mistakes like mine (see page 99). Bees have something like an internal GPS chip that orients them to home. It is very precise. When the more mature bees transition from being house bees and become foraging bees later in their life, they "calibrate" this chip by drifting slowly in front of the hive, their eyes fixed on home. This is called "orienting." Beekeepers all love to see this sweet curtain of new foraging bees drifting in lazy figure-eights in front of the hive.

But if something happens to the hive—say, the colony is in a tree that falls—the bees take notice of the abrupt change and reorient themselves to the face of the hive. Beekeepers can help bees relocate to a new site by placing branches in front of a moved hive, and even by rocking the hive gently as it is moved. This tells the bees something major has changed, and they need to reset their GPS when they leave the hive. Of course, there are always those who exit the hive in a rush to the flowers who don't take notice of the change. For these bees, it is helpful to place a bit of empty comb in a small cardboard box and leave it where the old hive had been. At dusk you will find the lost bees waiting patiently for you to carry the box to the new hive site where they can then be joined with their sisters. I know that sounds crazy, but they do it. Like a child lost in a store, the bees will remember the last place they were with their family and head to it.

This method of movement is what I did the second time that summer when I moved the Melissa bees. When I asked myself to start thinking like a bee, I realized Melissa needed better winter cover from our constant rain. I built her a small "bee shed" with four sturdy posts and a sheet of plywood for a protective cover. I sited it on the far side of the yard, well protected from the western sun and open to the morning warmth of the eastern sun. I have only primitive construction skills, so the shed, though not exactly a beauty, was a fantastic accomplishment for my bruised fingers.

An hour before dawn's early light, I placed a small screen across the entrance of the hive to lock the bees in. I chose a cool night so every bee was inside and there was sufficient ventilation with bees on the outside, as too much heat inside can cause a colony to suffocate. Then I secured the hive to a two-wheel dolly with car ratchet straps. Melissa and I bumped a bit across the yard, and I managed to lift the hive onto a level table under the new roof.

Then I tacked and tied branches to the front of the hive so the bees would not be able to simply fly out to forage. They would need to work their way around the many stems and leaves, indicating to them that something big had happened overnight. I pulled the screen off the entrance and, as the morning sun began to light the day, I was pleased to see many bees turning back to face the hive and drifting in lazy figure-eights before heading off to the flower beds. That night, I found 200 bees sitting on the comb in a box I'd placed where the old hive had stood, and shuttled them to their new home. Everyone else made it home safely on their own.

As the shadows of late summer lengthened, Melissa worked on her house while we worked on ours. Some of the work on our new place had to be redone, pushing our move-in date further out, and we were busy trying to set up service with contractors and medical providers. Being new in town and struggling with these new stressors, I found that sitting next to Melissa in the evening gave me unexpected peace and comfort. Night after night I found solace in the bees' quiet acceptance. But even in my sorry state, I began to notice that my bees were doing perhaps no better than I.

Melissa's numbers were clearly dropping. Some of the bees on her entry board looked as though their wings had been burned and shriveled. If you notice this happening like I did, you might do a quick internet search for "bees with shriveled, burnt wings." There, you will learn about a contagion called Deformed Wing Virus (DWV). Bees contract it from the Varroa mite, an unwelcome invader that moved into our country in the early 2000s and set about decimating bees by the hundreds of millions. The mite carries several nasty viruses it can pass on to bees.

Looking back at this time after our church hive cutout project, I understand now that cutouts are a brutal way to obtain bees. They are unprepared for the extent of upheaval. Often the queen is rushed into hiding by the maidens and the rest of the hive is taken without her being found. The cutout bees need to reconstruct an entire home in their new location quickly if they are to survive. It takes them a long time to rally and repair combs. If the queen is missing, they may try to craft and mate a new queen but that is not always possible. Without a queen, the colony dies.

On top of their traumatic removal from the eave of the church, Melissa had to survive my many blunders as well: being moved three times; the honey robbing as they set up housekeeping; the imprecision of the combs I tied into the hive. The bees spent weeks rehanging them and cleaning the floor of the hive where several had fallen. The stress on Melissa was enormous, allowing the mites and the viruses to take hold. A strong hive can face many challenges. But Melissa was not strong.

I sat by the hive in the low summer twilight, mourning the sickly, deformed-wing house bees who stood on the landing board ready to start their life as foragers, only to plummet to the ground where they wandered aimlessly, unable to fly and bring in food. These bees I would crush with my fingers, not wanting them to suffer any more than they already had. "It's been so hard for you," I whispered to the Melissa bees. "I am so, so sorry for all your suffering."

How easy it was to see what was happening to Melissa, and yet how hard for me to acknowledge what was happening to me. One morning in late October, I asked my husband John to drive me to the emergency hospital. I was struggling with depression and determined to see a mental health doctor. Very soon, I was admitted to the psychiatric ward and would spend a week at the hospital while my bees remained at home.

I arrived home late in the evening and was up at first light to hurry into the garden and see how Melissa had fared. I could see the hive from the backdoor porch, and my eyes flew to her entrance, searching for movement. There, someone was flying!

But my excitement plummeted quickly. Yes, there was flight, but it was not honeybees. In horror I watched yellow jackets flying in and out of the empty hive. Melissa had perished in the week I was away. My heart sank, and I stood with my hands gripping the sides of the hive. Though my heart hurt, I could not cry. The antidepressants that had pulled me out of my mental drain also left me as emotionally flat as a crepe.

In those silent moments of grief and regret for all the mistakes I had made with those sweet bees, I learned some things I keep close to my mind, close to my heart. First, it is not true that the gods never give us more than we can handle. And it is not true that what does not kill us just makes us stronger. Sometimes we do not emerge from calamity victorious. Sometimes the challenges are so great we do not emerge at all. Sometimes we emerge crippled and bent and a bit less able than before. But we have at least endured.

Sometimes just to endure must be enough. I had endured and it had to be enough. Melissa had not, but I would not let her short time with me be for nothing. In the years to come, I would make more mistakes with my bees, but I would never make the same mistake twice.

3

Building Confidence

"Can you tell me how large it is? Like, the size of an orange, or a football?"

"Actually, it's about the size of a tennis shoe." With those words the fellow over the phone gave me a pretty good image of the size of the small bee swarm he discovered that morning clinging to his rose bush.

"I'll be there in about a half-hour," I told him, and rushed off to grab my basket of bee gear. I had never collected a bee swarm before. I'd never even seen one up close, but I'd taken Jacqueline's classes and, in theory, knew what I needed to do. Jacqueline sent this call to me about the rose bush swarm. She loaned me a bee jacket and veil and gave me a fine pep talk that morning. "You know everything you need to know to do this," she said. "You just need some hands-on experience and today is the day for that."

Off I went, with the adrenaline of the unknown coursing through my blood. By the time I arrived at the small rental house in the heart of downtown Portland, my hands were shaking.

Do you ever have that experience of feeling like an imposter? Of thinking that anyone who so much as looks at you will realize you are out of your element? I rang the doorbell and watched myself try to sound like I knew what I was doing. He pointed to the rosebush I'd just walked past, and invited me to "Do your thing . . ."

And there they were, dangling quietly on a branch, like an old tennis shoe. The bees hung fully settled and very still. Tentatively, I cupped my hand and pressed it ever-so-slightly against the bottom of the swarm. They remained totally calm, their hum rising only slightly in curiosity to my warm hand. For a few sweet moments, they shared their warmth with me.

I quickly walked to my car to gather all my gear: swarm basket, jacket, and a small cotton bedsheet to lay under the rosebush in case I knocked loose any bees in my efforts to gather the swarm. Near the bush, I placed an empty cardboard box with a window of mesh screen taped on one side to provide ventilation. I opened my basket and poked through the contents, looking for my garden clippers.

My plan was to clip the branch the bees had collected on and slip it into the box. This is going to be easy, I told my tentative self. I reached for my jacket and then hesitated. It seemed a little extreme to go at this small bundle of bees with a head-to-toe hazmat bee suit. I left the jacket on the ground, knelt down in front of the swarm, and started talking to them.

"Bees, I'm going to bring you home with me if you will agree to come. I have a beautiful, safe place for you. And I don't know what I'm doing and could use your help in this." I raised up the clippers, snipped the branch, and placed the bees into the box. Somehow, whether it was my own amateur bee-whispering or a stroke of luck, not a single bee was shaken loose!

I packed my gear and left Portland feeling like quite the bee professional. I told the box of bees seat-belted into my back seat that I would call them Rose, an homage to the bush they settled on in Portland, the Rose City.

At home, three empty Warré hives were waiting beneath my small bee shed. Two were gifts from Jacqueline, and one I had purchased.

Over the years, Jacqueline and I have considered nearly every hive imaginable and numbered their pros and cons. We now have specific criteria that our hives must meet (see page 79), and our thoughts on Warré hives have evolved over time. But at this moment in my beekeeping career, I had settled on the Warré, a small, vertical wooden hive with the unique addition of a "quilt box" that sits atop the box the colony lives in, like a ceiling. It is filled with pine shavings or dried leaves, which absorb condensation and keep dampness from accumulating inside the hive.

Unlike conventional beehives, Warrés are more challenging to open and see inside. Conventional hives have frames of wood or plastic that are easy to remove and to look at individually, but this can disrupt the natural formation in which bees build their honeycomb. Warré beekeepers give the bees bars to append to, which they may or may not follow. In a Warré, the bees will take note of wind or drafts and defensibility to decide how the combs should be built, rather than letting the wooden hive determine the layout for them. For that reason, these beehives are frameless inside and the bees adhere the wax combs directly to the wooden sides of the hive. If you had to look at a single comb, it would be difficult. Warré beekeepers are more hands-off in their care than conventional beekeepers. They manage their hives by the box rather than comb by comb. The end result is more freedom for the bees and less tinkering required from the beekeeper.

Already from Jacqueline's teaching, I understood that it was better to keep your hands out of your beehive and let bees do what bees do. What I had not considered was my own temperament: I'm a woman who likes messing around with everything.

From the time I was very young, I wanted my hands on stuff: mud, worms, the dog's ears, the dirt beneath old logs, water. To me, a relationship with nature means very much a hands-on one. With a Warré, I needed to treat the square hive boxes as one "being." I couldn't pull out combs to look at individual bees, or the brood. Rather, I would have to learn by observation, which is a good skill to develop (see page 121). I was going to watch the bees coming and going from the hive, and study the behaviors they enacted on the landing board. Letting bees determine their own situation is a good way to beekeep. I've actually come full circle back again to loving Warré hives, but my first summer with my own bees in Warrés was a challenge.

When I got home with the Rose bees buzzing quietly in their carry box, I was shaking with excitement. This would be the very first swarm I ever tried to introduce to a hive, and I tingled with both anticipation and concern. When you are new to bees, being immersed in their world can cause a sort of bee amnesia. Now, what was I supposed to do with these bees?

"First, open up the hive." I reached back in my brain for my bee class teachings and Jacqueline's instructions. I removed the very top cover of the hive, then the quilt box with the shavings and burlap across the bottom. The wooden frames the bees would build upon were lined up like eight pages of a book.

"Gently open the top of the swarm box. The bees may be attached to the cover." I did, and they were. The swarm clung suspended in her entirety to the top of the box like a large, amber droplet of dark honey.

"Hold the box and lid above the hive, and give it a good shake. The bees will drop into the hive and immediately dive into the frames below..." I did. And they didn't. They did dribble down, but then they reversed direction, and poured up, like a geyser. A few hurried down into the depths of the box, but many leaped into the air, swirling around the hive and humming excitedly.

I was not the only one watching them in stunned, wide-eyed silence. Above me on the ceiling of my little shed was an umbrella wasp working hard on her tiny paper nest. A small cluster of bees floated up to her nest and grabbed hold of it. The wasp backed up in surprise at seeing her home suddenly filled with strangers. She made room for her guests, watching them warily from the back side of her nest, her yellow head and long antenna weaving from side to side.

A hundred thoughts shot through my mind at once: "This can't be right. Are they all going to leave? What should I do?"

There are lots of ways to invite bees into a hive. Over time, I would learn that there are easier ways. I kept my breathing calm and, once I got past the crazy frenzy of it, the bees indeed sorted it all out. They eventually converged and began moving into the Warré hive. I was so relieved. So was the umbrella wasp, who went back to tending her tiny nest, no worse for wear.

Once I closed the lid of the Rose hive, the bees were hidden from me, but for those I could see coming and going from the hive entrance. I had struck beginner beekeeper gold with Rose. She crafted her nest by the book. That is, she crafted her nest by the books I was reading on how bees behaved. Her ordered and active way of being brought me a sense of order and peace, as well. My own life has a paucity of routine, but Rose became my anchor that summer.

Within three days of moving into her new and empty nest, Rose's maiden bees had constructed enough comb to give the inside a sense of structure. The foraging bees began to bring in nectar and pollen for the colony. If Rose's queen started laying eggs within her first few days in the hive, it would be at least 21 days until the first hatch, followed by 16 days of in-hive maturation before I would ever lay eyes upon them. I marked a "maybe" date on my calendar. On that exact date, I watched in amazement as Rose sent clouds of young bees out into the flower fields.

Later in the month, I acquired two more colonies. I named them Frejya and Shanti, and placed them in a row beside Rose. I was able, then, to watch three utterly different colonies of bees react to the variables of the summer, and their own particular challenges. Rose was the litmus test I used for determining how well Frejya and Shanti were coming along with their new homes.

Alas, neither hive was functioning well, and I was too new to beekeeping to understand what I was seeing. Neither hive, no matter how hard they tried (and try, they did), was able to raise a successful queen. Painful as this is to watch, it is the risk each young colony makes when it starts a new home and is the way Nature sorts strength from weakness (you can read more about losing a queen on page 163).

By summer's end, Frejya and Shanti had shrunk down to a small handful or two of bees and I was frantic. I decided to add Frejya and Shanti to the Rose hive. Perhaps Rose could use more helpers going into winter. So, one fine sunny autumn morning, I took the lid off of Shanti. Two hundred bee faces peered up at me from the box below.

Carefully, as slowly as I could make my hands move, I sliced the sides of the comb away from the hive box with a special hive tool, and lifted the two combs that held the bees up into the sunlight. Then, I opened the door to a feeding box I'd installed in the Rose hive. I slipped the combs inside and shut the door.

Before I could overthink the choice I'd made, I opened up Freyja. She held far more bees, who also peered up at me expectantly. There was no way I was going to carve all these bees out of their box. Instead, I took the cover and the quilt box off of Rose, and placed the hive box containing Frejya on top. Then, I quickly replaced the quilt box and put the cover back on.

Rose began humming loudly the moment she discovered Shanti inside her feeder box. Then, when I added Frejya, the sound increased until the entire hive sounded like a herd of power mowers. I expected the worst. A few bees, far fewer than I feared, blasted out the entry of the hive and soared in dizzying circles around the hive. Six bees tussled with each other at the landing board. And that was it. Rose accommodated her failing sisters and life went on.

So, I ended that first summer down to one hive from three. I pondered all the things that had gone wrong in my yard that year. I wondered if the problems were because I was using Warré hives, but in truth, I didn't know enough then to make an educated determination as to what had failed.

Have you ever felt the awkwardness of learning to play an instrument? Holding the thing in your hands, on your lap, aside your chin, asking yourself how does this work? Your fingers can't find the right string, or hole, or keys. You concentrate so hard that beads of sweat drip from your brow. But then, as weeks and months pass, your hands, your head, and your heart become familiar with the feel of the instrument. Something inside wakes up and tells you to play this music, even if you don't yet do it well.

You will probably find this with your first few hives. With Rose, Frejya, and Shanti, I didn't yet have the memory of bees in my hands, heart, and head, but I wanted dearly to play with my bees. I told myself that the problems I'd had that summer were because of the hive style. I really wanted a type of hive that would allow me access inside so I could tinker and mess about with my bees. I wrapped and folded my thoughts around the idea that having better access to them would let me help them more.

I no longer think that tinkering about in a hive helps bees. What changed my thinking is what seems to change most folks' thinking: failure.

4

Sacred Swarms

I did encounter a source of great joy in the midst of some failures—my first swarm. A swarm means different things to different people. What is a celebratory spring event for bees (see page 68) can often be perceived differently. To production beekeepers, this natural process in the lifecycle of a colony is a loss of resources. Production beekeepers do everything they can to prevent swarming. They do not want to lose 20,000 bees who could be employed as honey producers.

The general public is frightened by a whirling dervish of buzzing chaotic stinger-laden bees. People are so afraid of this natural process that they run for cover and call the police or fire department to save them from the scary menacing danger.

But there is a third perception: Natural beekeepers like us see swarms as a much-welcome source of bees who happily populate our hives. When someone like Jacqueline or I collect a swarm in a place where people can watch nearby, we combine the gathering process with education and show folks the true wonder of swarming bees. Few people have ever had an opportunity to see bees up close, to learn how docile and outright friendly they are. Jacqueline has a photo of her collecting a swarm while 30 resting bees perched jauntily in a row on her hat brim. If bees can smile, they certainly were that day.

Rose hive was my first to make it alive through a tough cold and wet winter, and I was so excited at the possibility of seeing her swarm come spring. Entomologist Tom Seeley's book, *Honey Bee Democracy,* describes the complexity of swarm behavior, and how the bees prepare. I readied myself for Rose's swarm by studying every paragraph of Seeley's book so I could identify each step and action.

There are many indicators that a swarm is imminent, and soon I had checked off the boxes for them all. Bees often go very quiet before a swarm. They load up with a bellyful of honey, ready for their journey, and then take pause. With all work done, they set themselves down and relax with one last song together before the swarm departs for their new life.

Rose had gone silent. On the front face of her hive, a few thousand bees congregated in a swaying mass, a moving blanket, grooming themselves and each other to pass the time. Although I could not see into the hive, I know now that at the bottom of the hive, masses of bees clustered, waiting . . .

At 11 a.m. on a lovely sunny morning in early April, as I sat in the bee garden with Seeley's book, I noticed a sudden ramping up of energy on the face of the hive. Several dozen bees exited the hive and dashed through the mass of quiet bees. They blasted through the calmly clustered bees like tiny rockets, knocking bees sideways, bumping them awake. More bees leaped upon their sisters' backs and shook their shoulders, seeming to say, "Wake up, wake up! It's now!"

I sat with my nose inches from the hive entrance and I heard a sudden change in tone, as if the bees went from singing "Row, Row, Row Your Boat," to Handel's "Hallelujah Chorus" in three seconds.

And then it began. Bees launched into the air like tiny cannonballs firing from the hive. A flotilla (bip! bap! bup!) collided with my surprised face—instantly I backed up a few inches and watched the bees become a swarm. And come they did, roaring out of the hive in a jubilation of wings and song. Up, up they went until the entire yard was speckled with dapples of sunlight glittering through 20,000 wings.

I was so close I felt I intimately connected to the experience. Energy rushed up my spine as I memorized their frisky motions, the exuberant delight of each bee. Jubilation poured from every wingbeat. The very energy of a swarm is life-transforming.

Swarming bees are strongly disinclined to sting because that swarm will need every single bee to begin its new colony. A sting is a wasted life and, as we sometimes chant, "Every bee matters, every bee sings." When Jacqueline and I move swarms from tricky places like car fenders or a fencepost, we just cup them with our bare hands, so gentle are they in their swarm trance.

Rose's bees settled on a nearby forsythia branch. They settled down far quicker than I did. The excitement of having thousands of bees swirling around me, the skin-memory of ten thousand flittering wings moving the air over my arms and face intoxicated me with joy. My hands shook with excitement as I prepared the Warré hive that would hold this swarm. I named the colony Brigid, after an ancient Celtic goddess-priestess whose gifts include healing, springtime, and poetry. It seemed such a fitting title for this swarm.

Brigid's flight in the sun's rays would give Rose's mother queen a new burst of fertility. The brief journey to a new home, a home that didn't yet have a nursery where parasites could accumulate, would reduce the mite load in the mother hive by 70 percent. The left-behind colony would bring in fresh genetics through their newly mated queen. The new swarm would expand Rose's footprint in the neighborhood.

We call the first swarm out of a hive the Mother Swarm, because the queen mother, mated and ready to lay eggs, leaves with this first swarm. The hive left behind is gifted then to her daughters. The swarms that leave around a week or so after the Mother Swarm, carrying unmated virgin queen daughters, we call the Daughter swarms.

We believe we know a lot about swarming, yet there are so many questions: How do bees decide which bees fly off with the swarm and who stays behind? How do they judge how many bees are needed for each group? How does a colony decide how many swarms to cast in a season? When in their seasonal cycle do they make that determination? I'm fascinated by these questions to this day.

Since the bees came, we have dozens of stories that highlight each swarm's uniqueness:

The swarm we drummed down from the tall cedar;

The swarm I rescued by climbing aboard a huge boom truck and being lifted 15 feet in the air to a cluster suspended over my town's main street;

The swarm that wasn't gathered by dusk who, in a tizzy, sent a dozen fearful bees crawling up Jacqueline's jeans (all ended well);

The swarm who introduced me to my neighbors;

The swarm who landed on the entrance of a gated community and no homeowner would go in or out until our swarm crew arrived;

The swarm who on Memorial Day weekend moved into a new construction site's wall, where the workmen gave Jacqueline and her husband Joseph a standing ovation when they left with the bees.

So many tales to tell. Were there many swarms this year? Did they settle low to the ground? How many departed to high trees beyond my reach as I waved farewell from below?

I measure my summers in swarms, and I'll bet the bees do, too. Suppressing this primordial urge to reproduce seems completely wrong to Jacqueline and I. Why would we try to abolish a natural urge? What do you think that suppression does to a colony over the years? Nature finds it important, so we do, too.

One morning at Jacqueline's farm, I was making my way toward the back door when I heard that unmistakable deep thrumming sound. Instantly, my eyes scanned skyward. And there they were, coming over the roof like a giddy sprinkling of fairy dust. I yelled, "Swarm! Swarm!"

Jacqueline and I dashed outside right into a mass of twirling, dancing bees. Both of us were laughing, hands outstretched, happy bees looping around our arms, swooshing over our heads and around our bodies. The shining golden colony had found the empty log hive right in front of Jacqueline's kitchen window and claimed it as their new home.

As we stood in the wonder of them, they moved in a floating spiral to the entrance of the hive and funneled in. I will remember that day always. I remember the way the light followed the bees over the crest of the roof and backlit their wings and full amber honey bellies until each bee shone like a jewel. I remember the sound of them singing as they flew, their very voices a vowel in Nature's sacred tongue. I remember the bees landing on wisps of Jacqueline's blond hair, on my fingers and cheeks. And I remembered—again—how swarms have the ability to make time stand completely still.

Would we ever dream of suppressing such a celebration of life?

5

Better with Friends

In my third year of bee tending, I became a bit of a Goldilocks looking for my next hive. I was searching for a hive style that was "not too big . . . not too small . . . but just right."

Rose settled in for her winter rest while I got busy planning my coming bee year. I doubled the size of my shed cover and purchased three top bar hives. Jacqueline had some of her bees in top bar hives and it felt like the right time for me to learn more about this style.

"I like Warré hives," she said, "because they seem more natural, like a wild hive living in a log, but they aren't easily opened without causing an uproar and damaging comb." She explained to me that Warré management is pretty hands-off and that means the bees build where and how they like. If they want to change the ventilation inside, the bees just go ahead and create curved comb instead of straight, which makes it hard to lift a bar out. Each of the boxes in a Warré hive has a separate function—brood eggs and pollen in the nursery box, nectar and honey in the others. This can be great for the bees, but in an educational setting, you wouldn't be able to lift a single frame up to show students who is doing each task.

Top bar hives are not vertical (like Langstroths and Warrés, see page 84) but more like a tree that has fallen over and sits on a stand. A top bar hive is a long, narrow horizontal box. Strips of wood (the bars) are laid across the top so the bees build comb that hangs down from each bar. Bees work these hives from front to back instead of from top to bottom. Bees don't usually affix their combs to the sides of the top bar hive, so it is easier to gently remove a comb and have a good, close look at what the bees are doing.

This arrangement makes it easier to show new students what bees do inside their home. And it makes it easier for someone like myself, eager to tinker with her hive, to do so without disturbing the bees. I placed the three new top bar hives under my newly expanded bee shed and waited out the winter, eager to begin working more hands-on with my bees come spring.

Now, when I read my lengthy journal notes from my first year with the top bar hives, seeing the litany of my mistakes makes my teeth ache. I was lucky with swarms that year, and I collected enough to fill all my hives: two Warrés, three top bars, and a hollow log I fitted with a wooden cover.

That was, indeed, my big year of dabbling in my hives. Now I was a full-on "helper" to my bees. I combined weak hives, and added a queen to one who needed one. I turned one of my top bar hives into a "duplex" with a colony at each end, separated by a board.

I went into the hives a few times a month, looking for honey stores, brood, and new combs. In my journal, I question myself constantly, unsure of what I am seeing, or what it means. What constitutes a weak or a strong colony? How few bees can actually create a functioning hive after a long winter? During winter preparations, when bees die off in the fall, how many deaths are too many? How much honey is enough to carry them through winter?

And yet, all my tinkering in the hives had left me with many questions, no good answers, and a strong feeling that I was simply not doing the bees any good at all.

That summer, Shekinah was my most cherished hive. She had been gifted to me along with her top bar hive—an unexpected blessing—from a bee friend. Shekinah was six years old when she came to me. Six successful years with no hands-on management. She was primal and strong, her bees dark gray and beautiful. Her beauty and survival success had nothing to do with me, but she was my pride and joy.

That summer she cast several swarms of gentle, quiet bees. I felt a calm and centered joy seeing her bring each new swarm to fruition. If this continued, I could fill more hives than I ever imagined.

But come autumn, she began looking alarmingly like all my other colonies: she was failing. I went back over my notes, looking for a reason. What could have sapped the energy of such a strong hive? What had I done differently from her prior unmanaged years? And then I saw it. I had opened and entered her hive many, many times. The repercussions of my summer of hive-diving was showing me that I had introduced a stressor to this once robust colony.

Shekinah's daughter hive, sitting only feet away from her, had been slowing down for weeks, so I opened her hive and had another peek. Flipping slowly through the rear eight bars of comb, I was stunned to find no bees at all. Pele hive, Glory-Bee, Brigid, Rose, and Sophia, everyone the same. All the colonies were losing bees, and I didn't know whether the losses were tragic or normal.

Let me stress, and stress again: Bees are complex beyond our ken. In retrospect, every thought I had about my bees in these early years was incomplete at best, and just plain wrong at worst. Every decision and choice I made for them, I made in a vacuum of knowledge and understanding. I had hurried to place my bees in containers that allowed my ignorance to reign. As I sifted back through my bee journal that summer, I felt sick inside when realized that every single manipulation I'd done with my colonies set them back a step, and in some cases, proved lethal to them.

Only at this moment did I remember Jacqueline saying, "Every time you open a hive, no matter how careful you are, there will be some damage that the bees have to repair: Torn comb, dripping honey, a squashed bee who needs to be carried outside, cracked propolis that creates a draft, or even dispersing the healing hive air that helps keep them healthy. Each time we open a hive, a number of bees need to interrupt the task they've been doing and fix what you messed up."

That is what I had done, and one too many times. I pledged to myself to keep my itchy hands in my pockets and to question my puttering when I felt inclined to open them up yet again.

As we entered winter, I continued wringing my hands in concern for my bees. Hand-wringing is pretty much all you can do as a beekeeper in Pacific Northwest winters. Opening a hive in the cold season can be lethal for the bees because they don't have the resources or energy to seal themselves back inside. By March, only two of my hives remained alive: Brigid and Shekinah. Both were down to only a handful of bees.

Desperate, I foolishly tried placing a heating pad across the top bars of Shekinah's hive. It was supremely stupid and I never admitted it to Jacqueline, but that action demonstrated how far I was willing to go to save them. Bees are sensitive to electrical impulses, and the heating pad was a hopeless, last-ditch effort that had no chance at success. Alas, when there are too few bees to handle all the housekeeping and food gathering tasks, there is no bringing them back. They must have a queen, and enough bees to cover and warm any new brood.

This critical number of bees needed to keep a hive alive I only learned from years of watching variable clusters of bees working in hives. My question used to be "how many bees does it take to make a successful colony?" Today my question has shifted to "How many bees *in a particular hive style* does it take to make a successful colony?"

April arrived with her opulent blossoms, but I had no bees to meet them. Every hive had perished and my garden was achingly quiet. I knew full well I had helped their demise. Ashamed and grief-riddled, I knew that I would need to shift who and what kind of beekeeper I was if my future bees were to survive.

If your plea is urgent and compelling enough, you may hear the universe respond. Sometimes that response is a very quiet whisper, and if you are not listening closely enough, you can easily miss it.

Sometimes, though, the universe answers loud and clear. To my silent cry, "Please, don't let me kill any more of my bees," came this answer in the form of an email from Jacqueline:

"I'm working on a book and I am sending some of my friends a few chapters. I would really appreciate your feedback!" I admitted I did not know bees well back then, but I knew writing and editing, having done it professionally for many years.

My email back to Jacqueline was a long one, asking many questions: Who was the audience for this book? Is this a how-to, or more story-based? What kind of chapter arrangement was she thinking about?

Her response was quick and energetic: "Dear Susan, you are the only one who didn't write back and say, 'Looks good! Keep going!' I have plenty of material but I am not sure how to organize it. Might you be willing to work with me on this? What can I offer in compensation?"

And that began many months of days at the farm. I sat on the love-seat in Jacqueline's office as we worked through the chapters. She had years of written material. As a beginner beekeeper, it was easy for me to sort what information needed to come first so that any reader could make their way through the tangled wonder that is bee life. My not-secret compensation: I gained a broad and vast understanding of bees!

A book was written that spring over many pots of mint tea and many farm lunches: Jacqueline's beautiful tribute to honeybees, *Song of Increase*. And with the book came what we've come to call our "Bee Dialogues." At first, they were short conversations Jacqueline and I began as we rested between chapters, discussing what should come next. Most often it started with a question or an observation by one of us:

"Do you think queen piping is 'a call to battle?' Really?"

"Have you ever had a problem—I mean an actual *problem*—with ants in a hive?"

"I read that 80 percent of first-year swarms fail. Do you think they leave the hive knowing that their task is not to survive, but to set up housekeeping for the next swarm?"

And off we would go into long, deep discussions, rife with sudden insights and rhapsodies about our beloved bees. Over time, many of these bee dialogues became lessons in our classes, or fostered our further research.

As my friendship deepened with Jacqueline, we both discovered that we were no longer alone in our bee worlds. My first few summers with bees were spent mostly in my own head, and with my own hives. I had hesitated to call Jacqueline too often, but in the course of our Bee Dialogues, we each encountered a kindred spirit, and learned a powerful truth: Beekeeping is better with friends! Bees do nothing in isolation. They are all about the village. And our best bee tending is done when we have a small and dedicated bee-loving village to support us.

Humans, too, evolved in villages. We do best surrounded by a group of friends and family. I have about a half dozen good bee buddies now: people I can call on to help with a swarm, to carve a log hive with me, and to set up a bee-forage garden. Having friends and helpers magnifies the joy of being with bees.

Jacqueline had been my sounding board for my "hands-on" bee summer. Her advice had been consistent, and is consistent with our advice on page 101: Let them be. They know what they are doing. By the end of summer, over an icy glass of watermelon juice, I made the beyond obvious observation: "Every time I go into the colonies, I mess them up."

"Yup," she nodded. "It takes them two to four days to reset the hive temps and the propolis envelope. It impinges on their health, and you risk killing or injuring the queen, and chilling the brood."

"But how can I get proficient at hive tasks if I don't do them regularly?" I complained.

"Why do you think you need to get good at those tasks?" she asked.

She had me there. I felt the air go out of me like a punctured balloon. "Everything I do to help is of no help to them at all. So . . . even if I was good at 'hive manipulations,' the activity still does them no good. Is that what you're saying?"

"I'm not saying, I'm only asking. Do you think these activities—given what you now know about bees—really help them when you did the combining, the re-queening, the comb moving? Do you believe you are doing these tasks more swiftly and efficiently than they can?"

I paused for a moment, imagining the best possible outcome of all my bee fussing that year.

It was a sea change moment for me. "No," I replied soberly. It might have worked for me, but it wasn't what the bees would have done. And it was no good in the long run, either. I looked at my hands, and at my weathered, over-eager fingers. A sudden, poignant phrase came to me: "Hands-off, eyes-on." I use it all the time in bee class, now.

6

Steady Growth

"Yes, Susan, bees do best in hives where the outside temperature does not affect the inside hive temperature in quick spikes of heat or cold," Michael Thiele said.

This made so much sense. As I sat in Michael Joshin Thiele's workshop in 2016, my mind raced trying to absorb all I was hearing. Michael has worked with Jacqueline for years. The two of them were among the very first beekeepers in the USA who experimented with keeping bees in more bee-forward ways.

Michael founded an organization, Apis Arborea, that is dedicated to rewilding bees in trees. He makes log hives that he places in trees. He even studies his watersheds to find trees where wild bees already live deep in the forest. Michael is a Zen Buddhist whose spiritual life and bee life are deeply interwoven.

As he spoke, "aha!" moments came quicker than I could write them down. "Bees are beings with many expressions, and one is warmth," Michael said. Honeybees are unique. Unlike other pollinating insects, these communal insects maintain their familial roles and relationships through winter. Meanwhile, other species of native bees go to sleep and hibernate through the winter, tucked away in a safe cubby hole or old mouse den.

Over millions of years, honeybees evolved a host of mechanisms to regulate the temperature of their hives. However, much like humans living in an uninsulated house, there is a tremendous cost to a colony that lives in a thin-walled hive. To accommodate fluctuating temperatures through the year, the colony in a thin-walled hive continually eats significant amounts of honey to fuel the effort. In thick-walled tree cavities, Michael told us, bees eat far less honey over winter, because it doesn't take as much effort to maintain heat inside a naturally insulated tree cavity.

While other apiculturists and scientists were studying bees in man-made hives, Michael was studying their wild behavior in their ancestral nests: trees. Who are bees when left to themselves? Michael was an early teacher of the wonders of bees in trees and had collected observations known to few other beekeepers.

For example, Michael told us that bees in a tree have no need to form a tight ball, or cluster, to stay warm in the winter. Bees in trees move about freely in the well-insulated tree trunk, grooming each other, feeding their queen, and caring for the hive. In this calm and quiet setting, the need for fuel is low. Small, too, is the number of bees needed to survive the winter! There are a number of advantages hives have when they reside in trees, and we detail these in greater length on page 78.

Jacqueline echoed Michael's enthusiasm when I told her about tree bees. A few years earlier she brought a bee-filled tree to her farm. Jacqueline's tree hive came to her in late November when a tree was felled and the logger saw a cloud of disturbed bees rise up from the fallen log. She responded with her pickup truck and managed to cart home an 11-foot section of log with a fairly intact colony inside.

What had been lost in the crash was the shattered section containing all the bees' winter honey, leaving Jacqueline with the task of feeding the bees through the winter. "I was always surprised when I fed my log hive over the winter. The bees were always fully alert. When I opened the feeding door to give them another honeycomb, they sauntered up to collect the honey, no matter how cold the day. They were never in a cluster!" A decade later, bees still populate that tree, now set under a rooftop to protect the slow-decaying old log.

As you would expect, the workshop with Michael generated another series of Bee Dialogues, the most pertinent of which was our discussions about what sorts of hive styles were genuinely appropriate for bees.

This discussion has lasted several years, and has resulted in some wild hive innovations and observations (see page 82 for a breakdown of different hives). I still revel in the role of Goldilocks of the Hives, but now with more specific questions: "Too hot? Too cold? Too square? Too sterile? Too open? Too drippy?"

In my quest for a hive that was "just right," I kept up the search. It took a few years more, but I finally found it: my bliss is bees in hives made of straw.

I remember looking at a book about Sun Hives (a style of straw hive) and skeps (see page 82), wondering how I could ever learn to make such a thing. I'm a "show me" gal where handcrafts are concerned. Books describing how to weave leave me confused and fumbling. Michael came to the rescue when he decided to host the first Sun Hive making workshop in the United States.

You can purchase these beautiful hives in Europe, where they were designed, and there are many workshop opportunities to learn the weaving of them, but not here in America, where bees come in square boxes and that is that. I could hardly contain my eagerness to build one of my own.

Jacqueline and I talked constantly about what sort of hive (short of a log) would truly be beneficial for the bees. In my previous years, my focus was on hives that would allow me to do what I thought was needed by my bees. Now, with my new dedication to "eyes-on, hands-off," my concern was only what worked for the bees.

Jacqueline had issues with her top bar hives, and I, too, found them too damp for our rainy Pacific Northwest climate. Both of us busily explored better options. On an online bee forum called Biobees, beekeeper Phil Chandler had been exploring putting a specific kind of mulch on the bottom of his hives to replicate what might be found over time in the bottom of a tree cavity. He called this bottom ground an "eco-floor," and it made total sense to us.

The bees had shared with Jacqueline their need for what they termed "a fall-away," an area distantly below the combs where unneeded stuff could safely fall down and away from the colony. This fall-away functioned as a trash and recycling heap for the bees and a food pile for other little insects that made their home in the bottom of the tree hollow. In some hollows the fall-away even functioned as a tiny compost heap, providing them a bit of winter warmth.

In this eco-floor fall-away live all the symbionts and organisms that inhabit wild hives. These creatures, from fungal mycelium networks to microbial fauna, use all the trash the bees provide and turn it into a fully functioning bee village of incredible diversity (see page 88 for more on eco-floors). This important facet of bee life—the village beneath bees in the hive—is rendered impossible when we put bees in boxes. Since beekeepers started insisting bees live separate from their bug community, our bees have had to go it alone, something unfamiliar to them in 30 million years of life on Earth.

Soon another felled tree with a bee colony in it arrived on Jacqueline's farm, and she immersed herself into observing and imagining how different life was for bees in trees. What did trees provide that manufactured boxes could not?

I made arrangements to travel to Michael's Sun Hive workshop. As the workshop drew closer, I was more excited about this facet of bee care than anything I'd stumbled upon before. To prepare, I watched historical videos of bees being kept in straw hives in century-old European apiaries.

The more I read of the history of these woven bee basket-homes, the more convinced I became that this would be my path with bees: returning them to straw hives where they have dwelled for thousands of years. Cozy, heavily insulated (like a log!), small, breathable, able to hold many members of the bee village in its crevices and coils, and free!

By the time the course began, I was filled with equal measures of anticipatory glee and worrisome dread. I had put a lot of emotional energy and financial resources into my vision of straw hives. "What do I do if it turns out I'm lousy at weaving, or that it bores me to death? I'm so invested in this," I confided to Jacqueline.

"Well, once you learn this skill, the next project the bees have in mind for you will show up," she consoled me, "Besides, Susan, this is terribly exciting! Barely anyone is weaving skeps in the US, imagine if you could bring them back."

So off I went to Northern California wine country to learn how to make a hive designed by the venerable German Bee Master and artist, Guenther Mancke. Guenther sought to craft a beehive that would symbolically and physically resemble the whole bee. The hive is woven in two separate baskets and then joined together like a large egg. This is the shape of a colony that is free to build unconstrained. Bees build their combs in the shape of a human heart, and Guenther believed people should see this.

The hive is made to be suspended on high, encouraging us to look up to the bees, which Guenther believed is our right orientation to them. He encouraged us to look up, up in reverence to their wisdom and natural gifts. Then in his 90s, Guenther has since passed.

The workshop passed by in a flash, and I came home wondering how to continue on my own. In two short days, you can only cram in so much learning. My hands had yet to gain the memory of how to work the weaving grasses.

I managed to get the Sun Hive covered—or cloamed—with a thick coating of manure, clay, ash, and sand. This coating is good for bees, and also protects the straw hives from UV damage, their greatest threat next to rain. Once dry, the hive was suspended from cables in the largest part of my covered bee area, the Beethedral.

A couple weeks later, I invited a swarm of bees into Wing, the Sun Hive, and I officially began keeping bees in straw. That summer I spent my busy hours searching for weaving templates and binding cane. Wing's bees were thriving in their new home. Within weeks, gossamer clouds of young forager bees orbited the hive like shimmering planets. The Sun Hive that summer was the sun in my bee yard galaxy.

But while I loved the bees in the straw, my own style of bee tending didn't fit with the Sun Hive. I don't have the carpentry skills to build a special Sun Hive platform, so the hive was suspended as high as I could place it, which turned out to be the exact height for me to bump into with my forehead over and over again.

The interior components of the Sun Hive can only be built by very skilled woodworkers, with very particular wood—plywood that doesn't off-gas and is safe for bees. In Europe, they don't make their plywood with dangerous glues, but there was no way I was ordering plywood from Germany without needing to take a second mortgage out on my house.

What I learned in our searches for the "perfect hive" was that we wanted a smaller hive body, a volume that bees could easily fill in a short season if need be. Also, I began to understand the value of an eco-floor, and there was no place for an eco-floor in the Sun Hive.

I still hadn't located a source of weaving grass and I struggled with this conundrum until it was suddenly and unexpectedly overcome. One day, my husband John took a new route home after shopping. I stared out the window at an unfamiliar neighborhood, absorbed in a perfectly normal moment, when we passed a ball field. There, right in front of the ball field, was a long block of grasses! I shouted, "Ooooh! Pull over!" I leapt out of the car and dashed to my happy discovery: tall, smooth golden strands of perfect skep-weaving grass.

Historically, straw hives are woven from rye or other grain grasses. In my part of Washington, I could find nothing like that, so I started to explore whether native grasses might do. The Sun Hive class was two months past and I hadn't touched a strand of weaving grass since. I fingered the long stalks from the ballpark with their frilly seed heads, and as soon as I got home, I called the Camas City Landscape Manager.

Jim Gant was a bee lover, and he invited me to cut all I wanted of the Karl Foerster Feather Reed Grass. The arrival at my home of these beautiful cut grasses dovetailed with my decision to try and make a simple skep, something far less complex than the Sun Hive, and hopefully, something that would suit both the bees and me.

I don't believe myself to be a skilled craftswoman by any means. My hands are clumsy, and most of my projects turn out to be not perfect, but good enough. I started right in, trying to weave very substantial skeps whose coils were at least a full 2 inches thick. My intention was to better mimic the insulation of a log cavity. But a coil of straw that fat felt like a cranky python in my hands, fighting my every attempt to make the coil bend into a perfect circle.

By winter's end, I had woven my first three wonky skeps. Their artistry was . . . well . . . there was no artistry to them at all but I told myself the bees would not mind. As I wove, I hummed happy songs to myself, imagining my new way of being with bees come spring.

Thankfully, I found that weaving was not boring, and that somehow, I had the patience to stick with the process, which was a long one: each full round of coil length took me two hours to weave. After weaving for several years now, each coil still takes me two hours, but my hands have found the rhythm and the coils have become more symmetrical and beautiful.

Meanwhile Jacqueline was deep into understanding the lives of the colonies in her bee trees. She was delighted to see my skeps and was eager to learn how to make them as well. I spent a weekend teaching her the process, but she was nowhere near as fascinated with the weaving as I was. "This is just not me, but I love the concept," she said.

The idea of woven bee homes stayed in her imagination, though, and soon after she showed me a bee home she created. She found some beautifully woven baskets made by a weaving guild in Africa. One fit inside the other with 2 inches of space between the outer and inner baskets. She nestled them together, stuffed the space between densely with Llama fur, and—voila!—a skep! Untraditional, but workable, so she cloamed the outside and placed it in her outdoor gazebo and waited for the first swarm to arrive. The more we played with ideas for bee-friendly homes, the more we realized that with just a few guiding considerations (insulation, small size, eco-floors, etc.), the possibilities were endless. We outline these criteria on page 79 and are grateful our hive search has led us to greater clarity on how our bees thrive.

Other exciting things were also afoot in our bee world. Jacqueline handed the teaching of the beginner bee classes to me. After my first few well-received lessons, I saw that once the students left the class, they had nowhere to turn for support for our kind of beekeeping. The clubs in the area still leaned heavily on conventional methods and treatments, so we created the Preservation Bee Club.

Then, Heidi Hermann of the Natural Beekeeping Trust, an international organization that champions our kind of beekeeping, suggested to us that we form our own nonprofit. Heidi felt an organization would give us a larger and needed voice in the natural beekeeping community.

We started the paperwork necessary to form a 501.c3, and the Preservation Beekeeping Council (PBC) was born. Through our friendship, Jacqueline and I expanded our own bee wisdom through intense and frequent collaboration. With our classes and clubs, we attracted other like-minded bee lovers, and though our little nonprofit did not see the other side of 2020, we still relish the thriving beekeeper community we built. The brainstorming and innovations and commitment to deep and respectful observation and care of the bees in our yards rippled further into the world.

As I've already mentioned, bee tending and paradigm shifting are not activities to be undertaken alone. Slowly but surely, our bee village was growing: a small but vocal group of bee enthusiasts committed to learning and implementing what bees want.

Bees who build up too quickly in the spring are at risk of starvation if the weather does not provide a blush of spring flowers. We are aware of shifting weather these days, both politically and meteorologically. We are following the bees' example of taking things slowly and thoughtfully.

A small colony can react to their circumstances quickly, adding more foragers on a sunny day, or shifting to wax making when the rain rolls in; the bees come into their new roles in a matter of hours. And so, we worked to keep preservation beekeeping methods as flexible as possible and to take note of where this new community presence wanted to lead us.

We did not foresee that she would lead us to Holland—or that Holland would change everything.

7

Bee-Longing

I recognized him instantly from the Facebook photos we had shared. Ferry Shutzlaars stepped forward to help us with our luggage and I greeted him with a full-on bear hug. Dressed in his signature canvas vest and plaid shirt, Ferry smelled of pipe smoke and wood. The moment he hugged back, I was overcome with the familiarity of him. I felt I'd known him always, and that he had always been a teacher to me.

Jacqueline and I had just landed at Amsterdam airport, ready to attend the first international natural beekeeping conference. I had to keep pinching myself to make sure that we were really in the Netherlands.

When we first heard about the "Learning from the Bees" conference hosted by the Natural Beekeeping Trust, Jacqueline and I set our hearts on attending. We began crowdsourcing and found that friends, family, and folks we didn't even know were enthusiastically pouring dollars and pounds into our cause. By that time Jacqueline's book *Song of Increase* had been published the prior year and she was becoming increasingly known internationally. Jacqueline was invited to speak at the conference, and she heartily agreed. I tagged along to soak up as much information as I could.

To be sure we were not jet lagged before the conference, we arrived four days early, and were busy reserving hotels and transport when more good fortune plopped into our laps.

I was, by this time, deeply wrapped up in skep making and had already taught classes on making straw hives in my hometown. Even though I was the local expert, I was not remotely satisfied at my progress with making a good-looking skep. I'd found an online Facebook site called "Weaving Skep Beehives," and I was a frequent visitor posting photos and questions. The administrators on the site were Ferry Schutzlaars and Mike Albers, two wonderful men who took skeps and natural beekeeping to another level entirely.

Ferry had begun messaging me with generous tips on how to keep my bees healthy in such hives. Of course, Ferry was going to be at the conference, and I asked him if we could possibly meet up before the event so he could show me where my weaving was going wrong. When I told him Jacqueline, her husband Joseph, and I were coming a few days early, he promptly said, "Come to my place. You can stay here and I can show you some things. I'll come get you all at the airport."

Other than our online discussions, Ferry did not know me. I was humbled and thrilled by his offer. Ferry's home in Haarlem, Holland, was a three-story affair that he called "new" because it was only 130 years old. He had one hive in his small yard and his other bees lived on farm, forest, and garden lots elsewhere. His town was a place of river canals and huge windmills twirling enormous canvas blades to ferry water through the country.

The Holland conference was a keystone event, not just in shaping our trajectory with bees, but in changing our views of the world. In Holland, I sensed a stability, a maturity, a sensibility in the populace itself. Haarlem was lush with trees and flowers everywhere. People even cultivated flowers in the cracks of sidewalks. Organic markets were on every corner. There were far more bikes on the roads than cars. Waldorf schools were everywhere where children made colorful paintings, gave well-attended theatrical performances, and tended beehives. I felt instantly at peace among such a like-minded people, busily caring for their environment.

Ferry became our tour guide, eager to show us the bee world of Haarlem. His newest project was helping an artist turn her garden into a bee sanctuary, where she planned to invite schoolchildren to visit and learn about art and bees. We asked Ferry about the legalities of inviting the public to a private home. Who would they invite? Would they be strangers? If they got stung, who would be responsible?

He looked at us, surprised. "We put out fliers and children come," he said matter-of-factly. And what if the kids get stung? "Well, if the kids get stung, they learn to be more quiet and respectful around bees."

The garden was beautiful. Ferry had his skeps in a covered shed alongside top bar hives. At a Waldorf school nearby, we found a Freedom Hive, designed by Matt Somerville, and a large Sun Hive set on its own tall platform.

We'd seen pictures of all these hives, but never seen them up close, bursting with bees. For me, seeing beautiful skeps made by skilled hands and sitting in 21st-century bee gardens was akin to an audience with the Pope. The skeps were sacred icons of a different order of beekeeping. Every bee garden Ferry took us to was an outdoor cathedral of respectful homage to nature and to bees.

We visited a small bee museum and cafe set amidst extensive flowering gardens. We marveled at hives of every shape and material: skeps in the shapes of rockets and people's faces. Skeps made square with wooden frames inside. Wooden hives with side windows made to sit in your living room. Straw hives built in squares, wooden hives built like skeps.

"Susan, look. These hives are just like what we are doing!" Jacqueline exclaimed. All these alternative hive styles used the same guiding principles we were inventing on our own back in America: small space, heavy insulation, eco-floors, logs. Europe clearly had much more going on in the world of natural beekeeping than we had realized.

It occurred to me how long a history Europe has with its honeybees. The wisdom there is ancient and eclectic, driven by personal connections with bees. In post-colonization America, our history with bees is commercial. We didn't keep bees in barrels, clay pots, top bar hives, skeps, or logs for long. The Langstroth hive developed in the 1800s wiped out the brief bee history that came before, and few US beekeepers saw any reason to experiment beyond the status quo with their bee care.

After our tour of Dutch beekeeping, Ferry made time to sit with me and show me how to weave. I hung on his every word, absorbing in my heart, hands, and brain every twist and turn. I wanted to remember all of it. I wanted my hands to work like his, with easy tactile rhythm and flow.

By the time the conference began, I felt full with new bee information. I thought surely I must have learned all that Holland had to teach me. Jacqueline and I were already buzzing with ideas to take home and if our Holland adventure had stopped right there, I would have felt happily sated.

But we would discover that Haarlem and Ferry had been the first course of a very big feast. The table had been set. Ferry shuttled us all off to Doorn for "Learning from the Bees."

From all over the world, we had come: 300 people from 32 different countries. We shared a collective experience of being on the margins of beekeeping. We wanted a better way to be in relationship with bees, to find ways to raise bees without causing them harm. We were the renegades, the ones who broke tradition—a risky proposition, no matter what vocation or avocation you aspire to.

The conference center was located in a beautiful wood alive with the droning of insects and the call of birds. Jacqueline, her husband Joseph, and I rolled our luggage along a stone path, following dozens of bee lovers toward the large glass and wooden doors. This was our first time meeting many of our peers, new friends, and teachers.

The doors swung open and we heard a dozen languages sounding at once. On the walls and tables were bee photos, paintings, videos, and hives of every invention and style: round, tall, square, insulated inside with daub, plaster, and cork.

Jacqueline and I glanced around the room. "It's like being in a swarm!' she said with sparkling eyes.

The lobby looked like a Monet painting, awash with the splendid coloring of saris, headscarves, hats, shawls, and ponchos. Familiar names appeared on name tags: people I had corresponded with, or whose books I had read, whose videos I had seen, or who had shared their bee wisdom with me. Now there were faces attached to those names.

The next morning, Dr. Tom Seeley presented the keynote address on his near half-century of studying bees living wild in trees. He talked about how these wild bees of the Arnot Forest close to his Ithaca, New York home bounced back after being stricken with Varroa mites a decade before (see page 152). He outlined a series of practices that could be implemented in any bee yard to help bees survive better in our care. (His recommendations for "Darwinian Beekeeping" can be found in Appendix 2 on page 215).

I sat with Jacqueline for his presentation and we could not have been more delighted at what he had to say: through observation, trials and errors with our own hives, and then hundreds of hours immersed in our deep Bee Dialogues, we came to many of the same conclusions: Small hives? Yes. Swarming and letting queens mate naturally? Check. Encouraging propolis production? Absolutely! Superior insulation of hives? Of course. No Varroa treatments? We'd been treatment-free all along.

Seeley strongly validated all the work our own explorations had inspired. We were on the right track—Jacqueline's bee-directed path. The huge and unexpected blessing for us was that now we had hard evidence behind us. Tom revealed that the bees in his forest had undergone more than 600 changes to their DNA in only 30 years—changes that enabled them to survive and thrive alongside Varroa.

The truth is in the science, Tom informed us. Nothing can beat natural selection for creating a healthy, thriving bee. I whispered to Jacqueline, "I will *never* feel sheepish about the way I care for my bees ever again. Six hundred DNA changes!"

We immersed ourselves in panels and presentations, art installations and experiential processes. Every country represented seemed to have a brilliant project underway, or had managed to preserve valuable knowledge gleaned a thousand years ago. Bees have such a long and indigenous history in European and Asian countries where they have been close to people for thousands of years. The hives made to care for bees and the ways they are handled are so very different across the globe, most especially in developing nations where the old ways are still in use.

We filled our notebooks with new science, new and old ways to house bees, and new ways to think about bees and our relationship with them. It was fast and furious, each presentation uncannily linking with the last, each new observation greeted with applause and flurries of excited chatter.

During breaks, I would sit at Ferry's display table, where Mike and I began teaching skep-making to passersby. Every time I began weaving, conference attendees wandered over and asked to see what I was doing. One by one, each person asked me to teach them how to weave the coils and start a skep. Every time I returned to my little corner opposite Ferry's skep display, within minutes a small group gathered alongside me on the floor, holding little bundles of ryegrass in one hand and binding cane in the other.

Then a young German bee researcher, Torben Schiffer, stepped onto a panel with three presenters and completely blew me out of the water.

Never constrained by any desire to appease the production beekeeping community, Torben's work has been defined by his loving respect for bees, and by a strong desire to see bees kept in accordance with their needs, not ours. For the past decade, Torben collected the data needed to show, unequivocally, that our current style of production beekeeping in square boxes is deadly to bees. And he gathered strong data on what bees truly want.

With only 15 minutes to summarize a decade of work, Torben spoke swiftly about topics few beekeepers ever discuss: mold in hives; the importance of round space for bees; propolis and how bees create liquid medicine and sterilize hive air that reinforces their health and safety. He explained the environmental damage done by huge beehives that amass all the local nectar stores, depleting food sources for native bees. He even addressed the concept of eco-floors and book scorpions in natural hives (see page 88). The information poured out faster than I could take notes, framed in words steeped in honoring the bees. (Torben's research on these issues is presented in Appendix 1 on page 181).

The presentation ended and I pulled out my tablet, opened up a photo of my skep hives sitting on their log eco-floors, and snuck up behind the crowd who had gathered around Torben. Three rows back, I angled the laptop screen with my photo of my hives over my head where

Torben could see it.

He smiled and gave me a big thumbs up and a wide grin, "Yes! This is perfect!" he said.

That day at lunch, I saw Torben wandering with a tray and I called him over: "Bee Master, come share more!"

"Hey, Crazy Bee Lady!" he called out affectionately as he headed for our table. For the next hour, he shared with us his long and conflicted journey with bees.

"Well," said one of our table mates, "What you are suggesting is that I don't take much honey for a while and I am not sure I agree with that. The way I see it, I give my bees a good home and I provide good forage. To be fair, in return they owe me a bit of honey, right?"

Torben looked at her and was quiet for a long thoughtful moment. "The bees owe us nothing," he said softly. "We owe them everything. Leave them their honey."

From the start, the "Learning from the Bees" conference was really about learning from each other, and sharing what we each had been learning from the bees.

An old bee master from Germany told us that if we melted drone comb wax separately from maiden wax and made candles from each, we would see that they burned differently. Was this because the waxes are different, or is the difference in the bees who had grown in that comb? Ah, another mystery.

Every hour of the conference brought new knowledge, and we were beyond delighted. Jacqueline's presentations were well-attended, including a 7:00 a.m. presentation called "Beeing a Bee," a guided sensorial experience of the world from a bee's perspective. Expecting 40 people, we were astounded when 200 showed up with open hearts and minds.

The conference had a deep effect on us. Jacqueline came with a heavy heart, with high hopes and middling expectations. She was well aware that bees suffered from humans misunderstanding their relationship with bees. But she felt that even natural beekeeping was not enough to bring about the deep healing bees and humans required.

I came convinced I was a rank beginner with little to share with an international community, but my skeps proved that not to be the case. I had found my niche without even realizing how valuable a niche it was in terms of bee health and happiness.

We came expecting to learn more ways to teach new bee lovers how to keep bees. Instead, we found ourselves among like-minded and compassionate people who were co-creating a whole new way of envisioning our partnership with bees.

Holland was a turning point for us. What we brought home from the conference revised our methods of Preservation Beekeeping and put us completely out of the box of conventional—and even most natural—beekeeping.

Remember the bee. Remember her rubbing shoulders and tapping antennae excitedly with all her thousands of sisters, sharing information, offering help, doing good work. Remember that she does nothing alone, and that her community is her very life. Strive to be a generous, respectful member of her bien (see page 171). Your bees will show you with their success and joy that you, too, will never be beekeeping alone.

In Part 2, we share the "how-to" of preservation-style beekeeping. Every day, there are more and more people developing new, exciting, and invaluable practices and skills in being with bees. This is our way, and we share it happily and enthusiastically, because we have seen how bees thrive under this kind of care.

PART 2

Preservation Beekeeping:

Healthy Bees and a Kinder World

8

Understanding the Bee Family

Bees are one of those rare creations in nature known as superorganisms, which is defined by the Oxford Living Dictionary as "a group of organisms which behave in some respects like a single organism; a complex system consisting of a large number of organisms which itself behaves as if it were an organic whole, an ecosystem."

For our purposes here, we're going to keep this discussion of creatures who work together as one being to its most simple. Like ants and termites, a gathering of tens of thousands of related bees is akin to one organism. Each singular ant, termite, or bee is likened to a cell in a body. Each cell has its very specific tasks, and no cell is capable of life apart from the whole.

So tightly bound to the colony mind is each bee that she cannot live for more than a day or two without her hive mates. Each bee has her own mind, her own tasks, yet is moved to serve the entirety of the hive with the whole of her life.

When something affects the colony, the singular bees act not on their own behalf, but on behalf of the entirety of the hive. In this way, they will quickly sacrifice themselves for the safety of the colony. In the heaviest forage season of summer, they will literally work themselves to death to store enough food for their sisters and brothers.

We have come to see this "colony mindset" as a creative gesture of devotion and service, each bee to the whole. The life of the honeybee is not only of a different order from ours, but rather of a higher order: a way of being in full dedication to the "one being."

We tell our new beekeeping students that our first query as bee tenders is: "What do bees want?" We tell them that the ultimate query of the honeybee is: "How can we serve?"

Bees alleviate suffering wherever they encounter it, from assisting the mineral kingdom in mending the soil, to tugging a drowning sister from a birdbath, to helping the plant world to flourish. As we get to know bees better in these pages, we also will be asking the bees: "How can we serve *you?*"

There are many good books written about bee biology (see our Resources section on page 177). Mostly, we want prospective new tenders to know about how bees live. You don't need to know how many dental fillings your paramour has—you want to know who they are. So, let's talk about who bees are.

MEMBERS OF THE BEE FAMILY

Drones

Bees reproduce in two ways. One method is by swarming. The other is by sending out thousands of drone bees to carry the colony genetics out into the world. Hopefully, these drones will mate with virgin queens in the area, and the genetics of the hive will move further out into the world.

Drone bees, the males of the hive, are reared in spring, in anticipation of swarms and all the virgin queens who will need good mates.

We anticipate the arrival of swarms a week or two after the drones start flying.

These big-eyed, husky-bodied bees comprise 10 to 15 percent of the summer colony, and besides mating, they have some other interesting gifts. Unlike maiden bees who are vigorously prohibited from entering neighboring hives, a drone may visit any hive he chooses, and is always welcomed.

After drones hatch out in their parent hives, they often move in with unrelated hives quite a distance from their mother hive. This peripatetic existence of the drones ensures that virgin queens will have unrelated drones to mate with close to home if spring storms necessitate that the virgin queens mate in their own bee yards instead of flying out to traditional drone congregation/mating areas.

Maidens

Except for laying eggs (which she can do in a pinch if needed), maiden bees do all the work of a hive. A maiden moves through an enormous list of tasks in her short spring-summer life of around six weeks. By the time she takes her last foraging flights for nectar and pollen, she will have served first in the colony's nursery where she cleans, feeds, and tends the growing eggs and larva.

She then moves on to become a food storer, a wax builder, a propolis maker, a house cleaner, a guard bee, a scout, a groomer, shifting back and forth between tasks according to the colony's ever-changing needs.

As spring lengthens toward summer, the thriving colony is in full swing. Maidens tend the next generation of brood, construct comb from wax, clean and clean and clean some more, make numerous trips to gather water, and harvest the abundant flowers for pollen and nectar. They prepare and deliver the nourishing food to busy maidens, robust drones, and the enduring army of foragers, much of the food coming from glands in their own heads.

Queens

Nature has crafted the honeybee so that the success or failure of any colony rests on the shoulders of its solitary queen. Bees can lose any other member of the hive and make do. If they lose their queen and can't replace her, the colony is doomed. The life of a queen can be long. Some queens have been known to live up to eight years, although these days with pesticides and parasites challenging them, queens only live a few years at best.

It is the scent of this one bee that gives a colony its identity. The queen is everything to a colony. In my bee garden, my bees will not accept a new queen if they can't produce their own. They would rather perish than work for a "foreign dignitary" who is of no relation to them.

BEES BY SEASON

It is helpful to learn about bees, and the roles of drones, maidens, and queens, in the context of the seasons. Bees are so deeply connected to the rhythms of sun, moon, and climate that any of the choices we make for our bees need to be considered through the lens of the particular time of year. A simple way to remember bees by the seasons is to remember the numbers two and four.

Honeybees have two main tasks they devote themselves to each year, and four seasons in which to accomplish them: If they are healthy, they will prepare for swarming, a process that takes many weeks. Post swarm, their task is to requeen their hive and gather food for the winter.

Spring

Spring is the most active season for bees. Multiple and wondrous events are happening in hives all over the world. After a long winter rest, the queen bee begins laying eggs in earnest. Even though the flowers haven't yet budded, she lays so the colony will have plenty of late spring and summer foragers to collect flower nectars.

Each season has its risks and dangers for the bees. In spring, bees need to build up their hive numbers (which drop off steeply in the winter) so that they will have plenty of foragers for when the flowers begin blooming in earnest. If the bees encourage the queen to begin laying too early in the season, or to create too many new bees, a sudden onslaught of spring rains and cold can mean the starvation of the entire hive.

The bees must somehow intuit their numbers, the time of year, the honey they have remaining in storage, and the weather. All these three things need to dovetail with perfection if the hive is to survive this first hurdle of spring.

After a winter of dreaming inside their hives, bees gather again at watering holes, drinking deeply to clear their delicate digestive systems of the congestion and torpor of the cold season. At this fresh time of year, the hive makes ready for swarm season.

Everything about a hive of wild honeybees revolves around the year's swarms. Before the swarm issues, the bees have tasks to be done. Whether or not they attempt to birth a new colony, known as a swarm, depends on the strength and health of this colony in early spring. Was it a gentle winter? How many bees perished? What is the situation of honey in the hive—abundant or not? Is there any disease lurking in the bees? You might be able to tell they're readying for a swarm if they fall silent, almost like a calm before a storm.

Production beekeepers thwart swarming in a variety of ways. Some clip the queen's wings, preventing her from ever flying. Without a winged queen, bees cannot leave their colony. Some beekeepers create "artificial" swarms by splitting the colony into two or three hives and thus preventing the colony from having enough bees to stage a swarm. Some remove brood so the colony lacks the feeling of abundance. Some beekeepers stack empty boxes on the top of the hive so, again, the hive feels like it hasn't done enough work filling its own nest space. Emotional as it sounds, a hive that lacks the feeling of full accomplishment and exuberant abundance will not swarm.

We might see this in our own lives when we feel thwarted in some way: what happens when the environment removes joy, harmonious excitement, and exuberance? The net result is one of contraction and restriction, and sadness.

None of these gross artificial swarm manipulations beekeepers take can come close to replicating the finely tuned instrument—swarming—that natural selection has crafted to build new, healthy colonies.

Nearly half of the colony departs in a swarm and takes the old queen with them to find a new home. The mother colony—that is, the bees who remain behind—needs to grow and birth a new queen from eggs left by the departing hive mother. The queen bee left many queen cells behind, sometimes a dozen or more, when the swarm departed.

These new virgin queens in the mother hive have many tasks before them. First, they will hatch out and harden their insect exoskeletons preparing for flight. The maiden bees will sequester some of these queens away, and prepare the hive for several more swarms, called daughter swarms as these will be led by the virgin daughters of the queen.

When the bees decide to send more swarms (normally about 10 days after the Mother Swarm), the bees may encourage as many as four or five virgin queens to accompany the daughter swarm. When these daughter swarms reach their new homes, the remaining virgin queens will hurry out on the first sunny days to mate and return to the hive. In the end, it is the full colony that decides which single newly mated queen will reign. The rest are dispatched.

Daughter swarms that are not successful in getting their new virgin queens successfully mated will perish. It is the risk of swarms that the mother colony also may not survive the "birth." The swarm itself may fail to find a new nest and gather enough resources before winter.

If you have a few hives, even just two or three, you may find yourself surprised at the lusty activity of springtime and the bodacious spring swarms. In my bee garden, each colony sends out three or four swarms, one right after another over a few weeks.

The tasks of the bee tender in spring are gathering swarms, or working to attract swarms with bait hives (read more about this on page 114) This is the time we ready our empty hives to become homes for new bees, and watch closely the colonies that have swarmed and are working to establish their new queen.

Summer

The survival of the colony into the summer season requires that a healthy queen carry out her reproductive tasks. To check the viability of the queen, production beekeepers open up the hive and look through the frames in the nursery until they find the queen. If she is not found, her productivity is assessed by studying the brood comb to analyze her laying pattern.

In the style of Preservation Beekeeping, we let the queen reign over her nursery, unbothered by humans. We want the queen to reign over her nursery, unbothered by humans. We don't actually need to see her to know what is going on inside. We can see the colony's behavior at the front entrance and around the hive and get a very good idea what is going on inside—without butting in.

Without entering the hive at all, we can observe a number of helpful things during the summer:

1. Are the bees coming and going swiftly, with a look of purpose to them?
2. Are the numbers of bees coming and going over the weeks remaining steady?
3. Is there a lot of pollen coming into the hive each day?
4. Are you seeing orienting flights in front of the hive (clouds of new foragers weaving lazy-8s in front of the hive)?

These are the signs that a fertile queen is in residence. If you see otherwise—bees seeming less focused entering and exiting the hive, perhaps wandering aimlessly on the entrance. Observe closely and take note of other behavior. Does it seem that pollen collection suddenly slows down? This may be a sign that there are fewer babies being born, thus less need for the baby's food, pollen.

Are the bees cranky and defensive? That might be a signal that something is amiss inside and the disruption has every bee concerned. If activity around the hive drops off and you see fewer and fewer busy bees at the entry, they may have decided their queen has reached the end of her fecundity and is becoming less capable of laying the needed number of bees. In this case, the colony will divert effort away from brood rearing and arrange to create a new queen to replace the old one.

How does this happen? Maiden bees can create a queen bee from any fertilized egg, by crafting a special elongated cell of wax around the egg, and by feeding it exclusively on royal jelly, a substance the Maidens create in the glands of their heads.

But the Maidens do not select just any egg to crown. They seek out an egg of the "Royal Caste" by some means we do not yet understand. Scent? Size? Heat? Somehow the bees can identify what Cornell professor Dr. Tom Seeley has identified as Royal Caste eggs. Every hive contains a certain number of eggs from this rare genetic line the queen carries—eggs with all the qualities needed for a successful queen bee. When bees craft a queen to replace a failing or injured queen, the process is called a supersedure, and usually, the bees handle the task flawlessly.

Autumn

Swarming, queen rearing, food gathering, colony building: these are the spring and summer tasks of the bees. Then, with the onset of autumn, the work of the colony shifts. The blooms of summer give way to the blooming foods of fall: asters, goldenrod, squash blossoms, autumn sedums, and ivy. The light colors of spring and summer nectar give way to the darker, molasses-hued nectars of autumn.

Inside the colony, the queen begins laying eggs that will become the bees of winter. These bees, fed heavily on pollen stores, are called "fat bees," and they are able to survive all through the long months of winter when the queen will slow her egg laying to a standstill.

The air in the autumn bee yard is heavily scented with honey and nectar. Bees are scurrying to gather all the food stores possible, backfilling the hexagonal cells in the brood chambers with nectar. There is a tension in the bee garden, as the bees work hard to protect their winter stores from yellow jackets and robber bees looking to cash in on a quick source of honey already in the making.

Hummingbird feeders and fallen fruit are suddenly surrounded by clouds of hungry bees, looking for any sugar source they can find. In late summer, you may see bees fighting on their entry boards, working to defend against the interlopers. In this season, some hives completely succumb to yellow jacket attacks.

Now is the time to be sure you keep the entrances to the colonies as small as possible, allowing for only the passage of a bee or two at a time. Small entrances are much easier for the bees to defend.

This is also a season when you may see dead bees pile up outside the hives. Don't be alarmed. Each colony begins reducing its population in preparation for winter, keeping only the number of bees needed to heat the hive, to keep the queen safe, and to warm the brood that arrives after winter solstice. Each colony must make its own determination of how many bees to keep, weighing this number against the available honey stores, and how long they believe the winter will continue.

Honeybees are the only bees who maintain a hive all winter. All other bees, bumbles included, perish in the fall, leaving behind pupae or—in the case of bumblebees—queen bees that must hibernate in the ground over winter. All wasps die off now, after their last raids on honeybee colonies, sending queens off to hibernate like the bumbles.

The nights grow colder and longer. The precious sun that calls bees to their summer joy is speaking quietly now, hiding her face behind rain-fattened clouds, dipping ever lower in the sky. The bee activity in the garden slows, ceasing temporarily on certain chill days, then finally stopping altogether.

Winter

We say goodbye to our bees as winter settles in and the bees gather together for their dreaming time. We will see them only occasionally on days when the frigid winter wind hushes momentarily and the sun ushers in a surprise breath of warmth. The bees emerge for a quick cleansing flight, a poop run, out and back before the cold gets to them. They will not defecate in their hives.

Bees need to have a clear opening to the outdoors all winter, expressly for these necessary cleansing flights. In winter, bee tenders have the responsibility to check on their colonies to be sure the doorway is always unblocked. If you get snowy winters, drifts can cover the opening. A colony can suffocate from lack of air. If they can't get out for a cleansing flight, it may cause dysentery and death. If a number of dead bees clog up the entrance—as has happened to both Jacqueline and me—the bees can't find a path to the door. Make all your hive entrances an inch or more from the bottom so a buildup of dead bees during winter will not block the entrance.

We who keep company with bees feel their absence keenly during these months. Gone is the comforting hum that follows us in our gardens. Gone, the little bee who alights on our wrist to clean her antenna as we sit digging in garden beds.

We keep our bees close in winter by cleaning or making new hives and teaching classes on tending bees. It is heartwarming to tiptoe through the bee garden and place your ear to the side of the hives, listening for that soft, low "snore."

New bee tenders can get impatient during these long cold months, wondering if their bees are still alive. The urge can be strong to just take a peek inside the hive and see how the bees are doing. But this action may well mean death to the colony: once the hive is opened, and the cold air rushes in, the bees cannot restore the warmth of the hive. They will freeze to death.

Bees in winter torpor can look dead to an untrained eye. Too many times over the years we have gotten calls from students who opened their winter hives and assumed the sleeping bees were all dead. In one case, a fellow brought the hive into his basement to remove all the honey stores, and next morning discovered thousands of hungry, confused bees flying around his basement.

In the dead of winter, spring can seem a long way off. But the warmer days will finally come and the first foraging bees of the season along with them. Jacqueline is fond of saying, "Bees dead in January are still dead in March. No hurry to look," and she is right.

With the coming of the winter solstice, the bees dreaming away their winter begin to dream of spring. Solstice is the traditional beginning of the bee year. Inside her colony, where she has spent the winter resting and eating, the queen or perhaps the entire colony will begin to stir. The first eggs will be laid, the first brood capped with wax and tended by the nurse bees. In our part of the world, the hazelnut catkins herald this event, and on warmer January days, you may see the first foragers out collecting pollen.

As winter fades out, the bee year has begun again, and the cycle of swarming and foraging resumes. Soon, the buds will fatten on trees, shoots will lance up through the ground, and the plants will call out to their friends the bees with the flower vocabulary of color and scent. For 100 million years, this relationship has been evolving between bee kingdom and plant kingdom.

This cycle is vital in understanding the bee family and what drives them. As I began my beekeeping career, it took trial and error for me to acquaint myself with the superorganism that is a honeybee colony. But gradually, I learned the bees' seasons and as the years pass, I appreciate their cyclical lifestyle more and more.

9

Choosing Hives that Help Bees Thrive

At the end of the 19th century, Reverend Lorenzo Langstroth created a beehive with square, removable wooden frames. It was a revolutionary moment in beekeeping. Suddenly, it was easy to look inside the hive, to remove honey, to stack up the hive boxes and create towers of bees. It also became easy to stack and move hives from place to place, and migratory beekeeping was born of this hive.

In short order, most states made this hive a requirement. If you wanted bees, you had to keep them in "Langs." Many states had bee inspectors, and the Langs were easy for them to examine. Suddenly extinct were the old woven skeps and the hollowed-log hives of ages past. The Langs were crafted with the ease and efficiency of the beekeeper and the inspector in mind, and beekeeping changed overnight and forever with the advent of this hive style.

But no one thought to ask the bees what they wanted. And bees have been struggling ever since.

Nowadays, with the huge upsurge in hobby beekeeping, bee lovers again have a choice. Many bee tenders are beginning to ask what

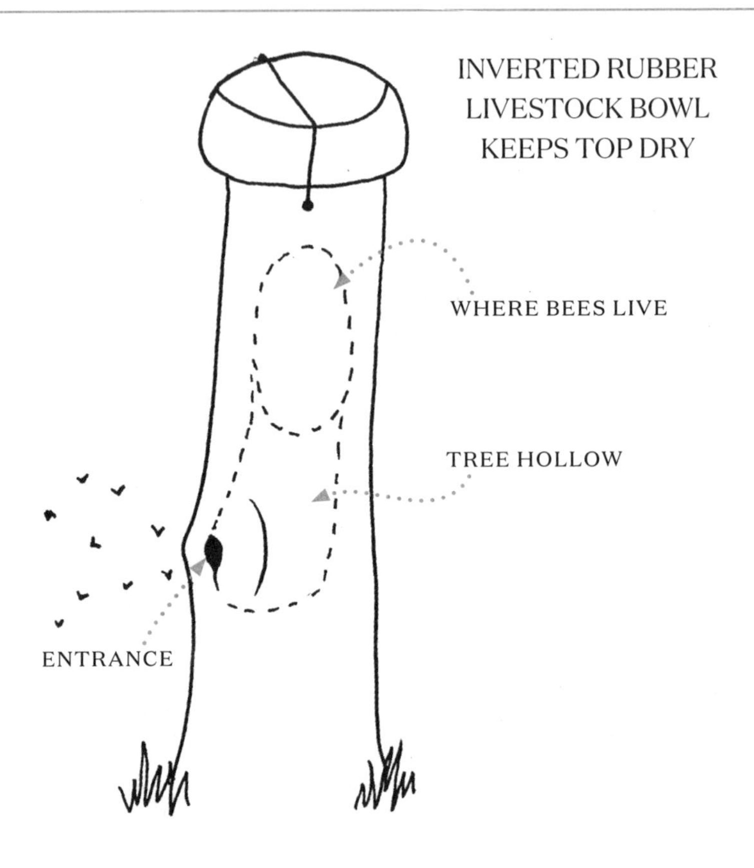

bees might prefer in a hive. The market has responded, and suddenly, there are a wealth of different hive styles to choose from: plastic hives, Flow Hives, top bars, Layens hives, Warré's, horizontal Langs, logs, and other new innovations hitting the market every month.

The choices are many, yet with our guiding question, "what do bees want?" it becomes easy to sort the options. Bees want trees, or tree-like habitat boxes.

BEES IN TREES

For millions of years, long before we started putting them in hives that suited us, honeybees have lived and evolved in tree cavities. So, if we are going to take bees out of their ancestral home and invite them to live with us, how can we best mimic all the benefits of the tree? Since our early struggles with our own hives, we have studied our bees, observing where they excel and listening to our peers' expertise on optimal hives. In doing so, our criteria for hives became:

- Heavily insulated so bees could maintain their desired temperature summer and winter
- Warm, dry ceiling
- Small size, round interior (around 30 liters)
- Low entrance, with a round tunnel to deter yellow jackets
- Eco-floor

Our reasons for these criteria are quite nuanced, based on the many advantages of log life for bees. Inside the womb of a tree, bees are protected by many inches of moisture-wicking wood. Bees want to live where the outside temperature does not affect the colony inside, and the thick walls of a tree enable the bees to set their own interior thermostat as needed. Bees heat their nests by shivering, keeping the brood nest area at a near-perfect 93° to 96°F—the temperatures needed by the developing brood—and the thick walls retain the heat the bees create.

Shivering requires lots of energy, and energy in the bees' nest translates into honey consumed. Bees in poorly insulated hive bodies use up to seven to ten times more honey than bees in trees. A well-insulated hive means that bees can survive the winter months on much less stored honey. It also means that the bees don't have to spend as much of their energy on generating heat to keep everyone alive. In well-insulated thick-walled hives, bees can spend their time communing all winter, warm and awake. With no springtime tasks like foraging, brood-rearing, and nectar processing to keep them busy, warm winter bees have plenty of social time to spend grooming each other. This behavioral difference encourages bees to continue grooming more year-round, too, and be less bothered by predatory Varroa mites. Thus, good year-round insulation carries many benefits beyond stabilizing temperature, including allowing even small colonies to overwinter successfully.

In the cozy environment of a well-insulated hive, a dome of warmth is created and maintained at the top of the tree cavity. The heat cannot wick out, because the entire body of the tree is above it. Moisture is wicked away by the wood fibers exposed at the top and bottom of the cavity, thus bees in trees are blessedly unaffected by the winter-killing effects of condensation and high humidity (moisture created in winter by the respiration of the bees).

The rough walls of a hollowed-out tree cavity encourage the bees to coat the nest with propolis. In the proper hive nest, bees use propolis and warmth to create sterile air in the hive. With this hygienic seal in place, pathogens entering the hive are immediately neutralized. In this way, propolis enhances the nest's ability to wick out moisture and provides the colony with her external immune system, which is activated when bees inhale or brush against the propolis, or when they drink moisture that has collected on the water-proof propolis.

Inside the tree cavity, bees build their combs in accordance with the air movement outside the hive. Combs are crafted in curves and waves that help the bees move incoming air all through and back out the nest. Bees are master engineers when it comes to heating and cooling their hives, and they do this by way of wax placement and vigorous fanning of their wings. In this way, the air in the hive is kept healthy, warm, and fresh.

Tree cavities are generally smaller than the hives we now provide for bees. Large stacked hives are not suited to bees. They are harder to heat in winter and require a large number of bees for the cluster. Natural wild hives maintain small, efficient nests, and can easily weather lean forage years.

Here in the United States, where honeybees are not indigenous, we must also consider the needs of our thousands of native bees. Gigantic stacks of honeybee hives deplete the forage sources in any area. Small hives allow for healthy populations of both native pollinators and honeybees. Small hives take fewer bees to make a thriving colony, and less honey is needed for winter stores. The small interior hive space encourages frequent swarming, and swarming is one of the ways bees keep themselves healthy and reduce mite populations in their colonies. And small colonies are much better suited to times of scarcity.

While we tend to think bigger is better, and imagine that a huge colony of bees is "stronger" than a small colony, bees themselves are very frugal, always planning for the seasons when abundance fails. A small hive is able to survive a downturn in weather and forage, where a large hive may crash.

Another problem with man-made hives is that these pre-made constructions are square. Nature is not a lover of corners, and neither are bees. Corners in hives are hard to heat and cool, and tend to become cold traps, mold traps, and hiding places for predators like hive beetles. In Holland, Torben showed us images of square hives, all with water damage and mold in the corners. Bees are sickened by certain molds and bacteria, especially those that form close to the ground. As arboreal creatures, bees have never adapted immunity to ground molds and pathogens. Bees exposed to molds are weak, sick, dying bees.

Most tree cavities are high off the ground, where bees remain nearly invisible in the high leaf canopies. In addition to the high placement, bees can also be strategic about the architecture of their hives, often setting up the entrance at a lower point to help with heat storage in the winter. Heat always rises to the top, so this low entrance placement can help regulate the bees' temperature by preventing warmth from escaping. When heat is trapped properly in this way, the bees don't have to work as hard to generate their own temperature with their body heat, which would deplete their energy.

At the bottom of the tree cavity, you would also find a population of beneficial insects and organisms—up to 8,000 different creatures and fungi that work with the bees to clean away old wax, dead bees, and create a dynamic compost beneath. This compost layer is called an eco-floor, and is often missing from man-made hives. A beehive truly is a village!

THINK LIKE A LOG: CHOOSING YOUR HIVE

Armed with this information, it's time to choose a hive that works for your bees. If you already have pre-made hives, there are simple modifications you can make so that your wooden hives are more log-like, and we'll show you how below.

But if you are starting from scratch, your options are unlimited. Honestly, just about any wooden container can be made into a serviceable, bee-friendly hive. Here, we've detailed the pros and cons of different hive types so you can select the optimal one.

Types of Hives

Skep: The skep hive is a circle, but even more than that, she is constructed by weaving in one, long, coiling spiral. Circle and spiral come together to make a creation of sacred geometry for bees. While these hives are my favorite, they have both pros and cons. The pros include size—small and round, which bees like. They are also naturally insulated and dry, with the round dome top where bees can make their tower of warmth, scent, and air sterility. The bees propolize them heavily. They are light and easy to carry. They are not just bee nests, but true works of art.

Cons include the time it takes to craft one. Unless you live in Europe, you will not find these in the states for purchase (Actually, China sells one, and it is thin, and fumigated with insecticides before it can be sent overseas. Avoid these hives like the plague). First, weaving grasses must be sourced, cut, and dried. Then, the grass needs to be raked or combed to prepare it for the weaving. Before the weaving starts, the grass must all be moistened and beaten with a mallet to soften it. Then the weaving begins. I can complete a full coil in about one to two hours. I occasionally sell one in the years when I have time to make new ones, and I sell them as artworks, because that is what they are to me.

WOVEN SKEP
WITH ECO-BOX UNDERNEATH

My master weaver friends can craft a skep in a day or so. It takes me weeks or months, as they are a lot of work. In the classes I've given on skep making, only a handful of my students have ever completed theirs. Also, the skeps need to be kept under cover of some kind. They will not survive out in a field or propped into a tree. I love this way of beekeeping, but you need to be a zealot if you want to do skeps.

Sun Hive: These hives are woven just like skeps, but are larger, and require a stout wooden frame of some sort for hanging. I find them a bit larger than needed in the size, and very hard to manage on their wooden towers or dangling from ceilings. I made one. Only one.

Log: These are constructed by hand from wooden log rounds. You can find these easily at small log yards, or around the neighborhood as trees are cut. You can slowly burn the core of the log out, you can try chain saws if you are gifted that way, but all the chain oil will need to be either burned or sand-papered out when your hole is done. There are also gouges made that you can use to slice away all the interior wood. Log hives are heavy, and need a very thick and sealed slab of wood atop. By thick, I mean a foot or so. That wood will wick and release moisture as the bees need, but the top MUST be sealed or the bees will be destroyed with the cold and moisture dripping from that upper cover.

Here are a few of the more standard hives that you can buy online. All would require serious innovations to make them truly friendly to bees:

Langstroth: These most common hives are designed for the ease of beekeepers, not bees. Their removable frames make it easy to look through the colony like a book, reading each comb. And that is their benefit: to the keeper, not the bee. They have no benefit to bees: bees cannot communicate across wooden frames that dangle unattached to the sides of the hive. They cannot heat nor cool in a hive with frames. The hive body is thin and uninsulated. It has corners that encourage deadly molds. I was given some of these, and I use them as planter boxes for herbs.

Warrés: These hives are smaller, which bees prefer. They have a quilt box above the bees where moisture is wicked away and condensation does not fall on the heads of the bees. Frames have been replaced with simple strips of wood across the top of the hive, so bees may affix their combs to the sides of the hive, enhancing the ability to communicate, and to treat each comb as a single room they can heat or cool at will. But Warré hives also have corners that are mold-makers. And the hives are as thin as Langstroths with no extra insulation.

Top Bar: Top bar hives are horizontal rather than vertical. For the beekeeper, this makes them very easy on the back, as no heavy lifting is needed. They are easy and inexpensive to build. Downsides to this hive are that bees have a difficult time in cool climates moving air sideways. Bees have evolved with the physics of cool air falling, warm air rising.

Bees in a cold top bar hive can easily starve trying to reach their honey stores that are hiding around a few cold corners, because these hives force them to move side to side, rather than follow the warm air up. By spring, you may find all your bees dead in the front half of the hive, and all the full honeycombs moldy in the rear of the hive. These hive bodies tend to do better in warmer climates, where they were originally developed.

Polystyrene Hive: No. Just no. Not only does plastic not breathe, but it can actually sweat and generate extra moisture and mildew.

How Many Hives will You Keep?

As you determine the type of hive you use, the number of hives you keep will also be crucial in determining the health of the ecosystem for a few miles around you. So consider your decision carefully. Honeybees are such efficient pollinators by visiting only one flower per forage run, that they are able to deplete the nectar and fully pollinate the flower quickly. Once pollinated, the flower dies, its work of crafting seeds for the coming seasons all complete. Native bees gather pollen from many flowers in their forage runs, meaning that any flower will be pollinated with fewer of the pollen grains it needs. Thus, each flower lasts longer, and more bees will be able to take nourishment from it.

For this reason, we advise keeping only two or three hives in your yard. Urban beekeeping is becoming so popular worldwide, with little governance on the number of hives that may be kept in a yard. In Berlin, once the main forage season is over, thousands of bumblebees are found dead of starvation on roads and sidewalks. Clearly, this is the result of too little forage, and too many honeybees.

We want to see all bees thrive, so watch closely in your yards and neighborhoods for native bees foraging. You should be seeing a lot! Bumbles, sweat bees, carpenter bees, wool carders, leaf-cutting, long-horned bees—an abundance of these pollinators will help indicate that the honeybee numbers have not outstripped the nectar resources in your particular area.

INSULATION

There are many ways to insulate a hive. If you are good at woodworking, you can build a double-walled hive and stuff straw between the two layers. Adding wooden layers will make your hive considerably heavier, but we won't be stacking boxes or shuffling the hives around, so the weight is not so much of an issue.

While in Holland, we saw an insulation method that we found intriguing. Rather than place insulation of some sort on the outside of the hive box, a suitable round form was set into an empty hive box. Then a mixture of chopped straw, cork bits, and plaster of Paris was pressed into the voids on the outside of the round form. When the mixture dried, the form was removed and—perfecto!—you have a well-insulated box with no mold-trapping corners.

We use a version of this method now, using a wooden cylinder fashioned by one of our bee club members that fits a Lang hive, a Warré, or any other wooden box or barrel. The round wooden cylinder is placed inside an empty hive box, straw or wool or other bee-friendly insulation is packed all around, and you have a well-insulated bee home.

These cylinders can be purchased online (Barry Malmanger, referenced in our Resources on page 177, crafts these). Wooden bars for comb-building can be placed across the top of the cylinder for the bees.

If you are using old Warré boxes for hives with a cylinder lining, three boxes are plenty. If you are using old Lang deeps, you'll need only two.

It is vitally important to retrofit any vertical hive (or clay pot or barrel or box) with a moisture-absorbing top, and a box for beneficial organisms and creatures on the bottom. We call these hive additions quilt boxes and eco-floors and believe they are as necessary to bees as the hive body itself.

QUILT BOXES AND LOG SLABS

As we've stressed, a beehive and especially the dome area of a hive must be naturally warm and dry. During winter, the respiration of bees in their hive creates a great deal of moisture. During summer, the bees dehydrate the nectar to turn it into honey. As the nectar gives up its moisture, the ripening honey also fills the hive with heavy, wet air.

This water needs somewhere to go, which is why we want our hives constructed from natural materials that breathe and wick away the wet. Moisture often goes to the top of the hive where it condenses into cold water drops and falls down on the bees. Humidity past a certain point encourages the growth of molds, which are deadly harmful to bees.

Warré had the right idea when he placed a small box of shavings or old dried leaves on the top bars of his "People's Hive" to absorb moisture. You can recreate this innovation by hammering together a box with 4- to 6-inch sides. Tack a piece of cotton cloth or burlap across the bottom, and fill the box with unscented wood shavings.

Place this quilt box on top of the frames in your hive, adding another sheet of cotton or burlap between the box and hive. Warré advised mixing rye flour with boiling water to make a watery slurry to paint it onto the bottom of the burlap/cloth. Bees don't like the taste of the rye, and it's an easy way to keep bees from chewing through the cloth and raining wood shavings down into the hive. With this simple addition of a quilt box, you have solved the problem of condensation dripping onto the bees.

Even easier than a quilt box is a thick (10 or more inches) log slab placed as a lid to the hive. The top of the slab needs to be covered with rain-proofing material. This slab will very effectively wick water up and away from the bees.

ECO-FLOORS

Now, let's turn our attention to the bottom of the hive. The bottom of the hive must accommodate a village of symbionts and beneficial organisms. In nature, bees in trees like to have a bit of distance between the bottom of the comb and the place where everything that

CROSS-SECTION OF BEE TREE
SHOWING FALL-AWAY/ECO-FLOOR
AND DECAYING TREE MATERIAL

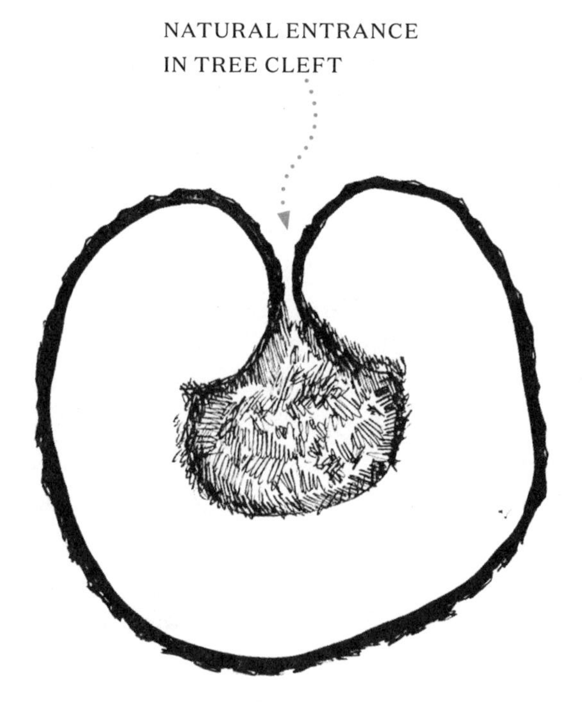

drops from the hive settles. Jacqueline calls this space the "fall-away." When a bee drops something, it falls down into a living, breathing world far beneath them.

In our bee gardens, we create our own version of the fall-away at the floor of a tree-hive cavity. For millions of years, bees have lived with a diverse village of organisms in an Eden-like, mutual admiration society. Ants, spiders, book scorpions, springtails, mites, mushrooms and fungi, beetles, moths, yeasts, and thousands more creatures made a dynamic world of heat and compost and pest management beneath the combs of the hive. It is estimated that more than 8,000 different organisms dwell with bees in tree hives.

Some of these creatures, like the book scorpion, actually ascend onto the combs to eat mites and groom the lice off of the bees. Tiny spiders prowl in the bark chunks and flakes of old comb. Earwigs and pill bugs eat mites and lice that fall from the bees.

When we moved bees out of trees and into our hives, we didn't make any accommodations for all those others. We broke a chain of relationships that had developed for mutual benefit over eons. Now is the time to compensate for this loss by restoring the eco-floor, a term first coined by British natural beekeeper Phil Chandler (biobees.com).

You can use an empty wooden hive body or a box to create an eco-floor. We also use hollow, 8-inch-high log rounds when we can find them. Tack a board across the bottom of the box or log round, and a board with an ample hole on the top (about a 5-inch diameter center hole), and fill the box with bark chunks, wood pieces, dried leaves, and grass clippings. Remove the bottom from your hive and place it atop the eco-floor where the bees may have full access to it. Over time the bees fully propolize the detritus in the eco-floor. If you leave the entire eco-floor entirely open to the colony above it, they will build down into the eco-floor, as well.

On many of our eco-floors, we fashion a small door on one side. If you need to feed your bees, it is simple to use this door to place a bowl at the bottom of the eco-floor. You can also hold your smartphone inside the eco-floor and take a photo of the bees working above. It's

ECO-BOX WITH REMOVABLE DOOR
FOR FEEDER OR TO ADD MORE
WOOD OR SHAVINGS

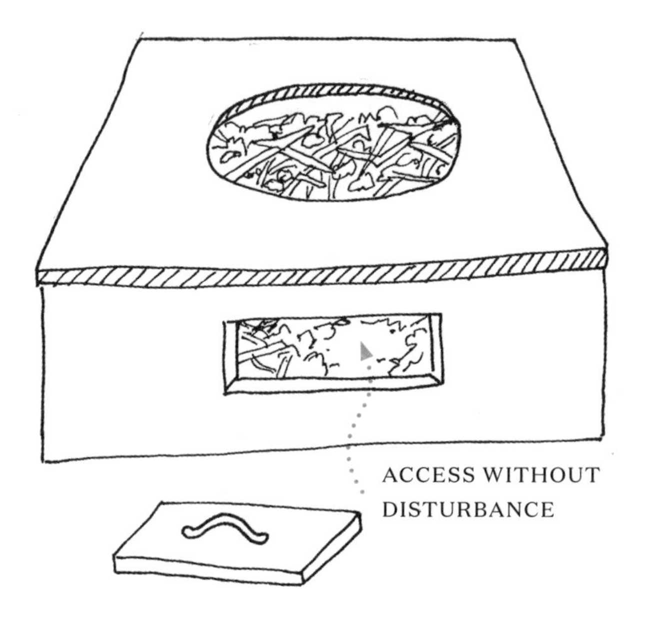

a great way to see how a new colony is building up without opening up the hive and causing a disturbance. While our straw hives have no need for a quilt box topper, they, too, benefit from being placed atop an eco-floor.

By sandwiching your hive between these two moisture-absorbing additions (quilt or log slab above, eco-floor below), you will come very close to mimicking the benefits of a log hive.

LOG HIVES

But what if you say, "Why bother with other pre-made hives? Can't I just make a hive from a log?" Yes, you can. Log hives are becoming popular as natural backyard hives. Hanging empty log hives on private and public lands to provide natural homes for bees is a process known as rewilding.

We both have bees in log hives in our bee yards. Jacqueline's came from sections of trees that were cut down and then discovered to contain bees. I have a big section of hollow log that I invited a swarm to inhabit. We both love the idea of bees setting up hive-keeping in their ancestral homes.

MATT SOMERVILLE'S FREEDOM HIVE THAT SITS 15 FEET HIGH

Once the logs are set in place, we keep them hands-off. If the hive perishes, the logs are quickly repopulated by wild bees, happy to have found such a great nest. If you search online for YouTube videos by the International Tree Beekeeping Movement (called "bee zeidlers"), Matt Somerville, or Michael Thiele, you will find some excellent instructions for making such hives.

The process of hollowing out a log with a chainsaw is not something every bee tender is prepared to undertake, but if you scout your neighborhoods, you may be surprised to find hollow logs tossed into a jumble of a neighbor's firewood pile. We have obtained over a dozen of these in the past couple of years, just by keeping our eyes open as we drive around the neighborhoods.

You don't have to be fancy. Two hollow rounds, stacked one atop the other, works. It's easier to transport two small hollow sections than one big heavy one and, from our experience, the bees find these prime nest spaces and move in on their own.

Add a quilt box to the top, or a thick log slab (sealed from moisture with wax or some other covering) and place mulch and bark chunks into the bottom of the rounds as an eco-floor.

CAN MY BEES JUST LIVE WITH ME?

Long before humans kept bees in hives, we kept them in our homes. We call these bee homes wall hives, and the earliest yet discovered was in an ancient stone house wall in Rome dating back to 800 AD.

If you think about it, a wall is a perfect, log-like place for bees. The in-wall hives are narrow and tall. There is thick insulation on all sides. Families would carve such spaces into their cobbled or wooden walls, making an entrance door either on the outside of the home, or inside, where honey could be gathered from the bees as needed. A small hole to the outdoors enables the bees to bring home the nectar and pollen into the house.

When Jacqueline and her husband bought their farm, bees came with it, already snugly ensconced in an uninsulated wall with an entrance under a chipped chimney brick. Every morning on her way downstairs, she swings open the window a few feet from their entrance and watches them fly in and out. In the cold of winter, when no bees are flying, she puts ear to wall to hear their hum from her hallway. Twice in 15 years the hive went briefly empty, but each time, the nest was claimed by a new swarm very quickly.

Last year, I had the space between two uninsulated studs in my bedroom opened up. We drilled a small hole for a bamboo tube through the house siding onto the patio, and covered the face of the cavity with double-pane glass. On the bottom, I fashioned a small door. Then, I introduced a swarm of bees. My bedroom wall hums with the music of the bees. The glass is covered with a black-out, insulated quilt.

Bees are lovely as roommates. Bringing bees into the walls requires special planning. Consider having the electrical outlets beneath the wall hive disconnected to help protect the bees from the electrical currents in the wires. Also, keep scented candles and perfumes out of the bees' room as bees are very sensitive to artificial odors.

The room's temperature must be kept constant so that the bees don't have to adjust their hive temps all the time. And keep noise in the room to a minimum. When viewing, do so at night with the room lights off, and a red cellophane cover or filter over your flashlight (which is a light source undetectable to bees).

While we enjoy the opportunity to live so closely with bees and to be able to peek at them occasionally, we never want this gift to come at a cost to the bees. Always, our guiding mantra remains, "what do the bees want and need?"

Reflections from the Hive

Susan

I've been dancing with depression for all of my adult life. Sometimes, depression and I waltz. Sometimes, I'm dragged around the dance floor in a death grip.

But I can honestly say, I've never been touched by a moment of depression, not the slightest whiff of it, in the presence of bees. In Europe, just the scent of bees and beehives are considered healing, and people pay money to breathe the air in apiaries where cots are placed in the center of a circle of hives. It is a common medical practice.

Because I have the delight of bees inside my bedroom wall—with a small open screen into that wall—I can dream in the scent of bees, propolis, and honey all night as I sleep, and listen to the murmur of a happy, thriving nest.

But also because I work around my bees with little-to-no gear, I must work especially slowly and gently so that the bees remain calm and happy with my hands among them. Bees do not like the scent of our breath, and they can become stingy if you breathe on them. So I quiet my breath and slow it down deeply. When you slow your breath and make it shallow, you activate the parts of your nervous system where peace, joy, comfort, and wellbeing live. This research is beautifully presented in the book "Breath," by James Nestor.

Slow, calm breath, slow movements, the intoxicating scent of bee nests, the joy of being "welcomed" by 60,000 winged ones: depression can't live here. No small wonder I spend as much time with my bees as my life allows!

A FEW FINAL TASKS

So, you've crafted a hive now, something cozy and dry. All of the hive ideas we've shared are vertical in orientation, which is best for bees in temperate climates, so you've got that guideline handled.

If you haven't crafted an entrance for your hive, now is the time. Bees prefer entrances toward the bottom of their colonies. By keeping the entrance low, the dome of warmth at the roof of the hive remains warm and dry.

But if you place the entrance too low, during winter dead bees may stack up on the floor of the hive and block the door, smothering the bees. About 2 inches up from the bottom of the hive, or even in the eco-floor beneath, drill a 1- to 1.5-inch hole, and find a 3-inch length of bamboo or even PVC. Our entrances are made from short 4-inch lengths of bamboo that are about 1 inch wide. That means an invader has to run through a tunnel alongside and surrounded by all the hive bees. It is quite a gauntlet! This tube entrance mimics a thick-walled tree hive entrance and provides an extra layer of protection. Push this through your drilled hole, forming what we call a "bee gauntlet." This kind of entrance protects bees from robbing, unless the hive is very, very weak.

And of course, the right hive you have so beautifully created will be even better when it is placed in the perfect location for your bees. Let's go find that spot.

10

The Right Location for Your Bees

We can help bees flourish by making wise decisions about where to place hives. Bees are creatures of place and prefer to stay in the same spot. As I learned the hard way, it is not impossible to move your bees, but it takes many days for the bees to regroup themselves and adjust to the new location. Here are a few guidelines for hive placement. With these in mind, you can choose the best spot your garden has to offer:

- Place your hive so it has some midday shade.
- Winter sun is good.
- Place hives as high off the ground as practical.
- Ensure winter cover/shelter of some kind.
- Place multiple hives some distance from each other.
- When multiple hives are nearby, aim the entrances in different directions to minimize bees drifting into neighbor hives.
- Remember that bees fly up when leaving the hive.
- Face the entrances east or south.

Bees in trees are usually high up in a sheltered canopy. Sun beating down on a hive all day can be harsh, and bees will devote precious resources to keeping the interior cooled. Bees would never choose to set up a nest on the ground in the middle of an unprotected pasture, which is where you see many hives these days.

Take note of the hottest times of the day in your climate. Here in the Pacific Northwest, our hottest hours are between 4:00 and 6:00 in the afternoon, and our western sun is brutal. We know how the sun moves across our gardens and make sure no hive is bombarded by western sun. Can you find a place on your land where the hive is shady during the heat of the day? If not, perhaps a wide roof or patio will keep the sun at bay.

Winter sun is always a good thing, and if you are lucky enough to have deciduous trees and vines in your yard, find a spot with dappled shade in summer and some direct sunshine come winter.

If hives are placed on or close to the ground where ground molds, bacteria, winter-wet soils and fungus thrive, bees will suffer. By nature, they are unadapted to these problems. Keeping hives at least 3 feet off the ground enables you to work around them more easily, and keeps the bees away from the soggiest aspects of winter ground.

We set our hives on picnic tables, on stools, on old cabinets and bookshelves, all protected by some sort of cover, such as a shed, large waterproof umbrella, a roofed porch or patio, or under a big eave. You can even place an old, waterproof campaign sign on top of a hive and weigh it down with a rock. No matter how stout and waterproof your hive seems, putting them under cover is extra insurance for keeping moisture out of the hive.

Traditionally, hives are set with their entrances facing east or south so the morning's sun warms them early and they head out quickly to forage. While this makes sense, we've never noticed that bees much care which direction the hive faces.

If you have neighbors close by, direct your bees away from your neighbor's yard. Remember that bees exiting the hive will almost immediately aim themselves upward toward the sky and not straight out. The best idea is to keep them a good distance from a neighbor's yard, but if the best location is near the property line, you can steer bees up and away with a tall potted plant or a piece of wooden lattice screen a few feet out from the entrance of the hive. Your neighbors will never see the bees coming and going.

WHEN A HIVE NEEDS TO BE MOVED

If you realize you've made an error in choosing the best place for your hive and find you need to move it, ignore the old beekeeping adage that tells you that you must move the hive only 3 feet a day, or move it 3 miles away. Accept that anytime you move a hive, there will be some calamity.

If you move your hive miles away to another yard (some beekeepers keep their bees in several different yards, or with neighbors), or even just 10 feet across the yard, confine the bees in the hive the night before by plugging the entrance with a mesh screen. Only use a screen! Bees breathe a lot of air in a short time; a solid plug of any kind could quickly suffocate the bees inside.

Come morning, carefully relocate the hive. Bees moved a distance of miles will not attempt to head back to their old homesite, and can have the entry-blocking screen removed as soon as they are set in their new surroundings.

But this is not true for bees moved within your yard. Clip some branches from nearby shrubs or trees. We're not talking little twigs here, or snippets of grass. We mean *branches*. You must convince the bees that something drastic has happened overnight, enough so that they will consciously take note of the change and reset their inner GPS chips for this new hive location.

By moving the hive, jiggling it a bit as moving naturally does, then placing a bunch of branches across the entrance forcing the bees to literally crawl their way out of the hive, you will encourage most of the bees to reorient. Those that don't will come back to the old location for several days. Leave a small empty hive box, or even a cardboard box with some honeycombs inside, for the lost bees at the old location. In the evening when they have come home to rest, you can take the box and gently shake it out by the entrance of the mother hive, and the bees will find their way back to their sisters.

CLOSE BY

We delight in having at least some of our hives close by. It is wonderful—and helpful—to be able to hear and see your bees as you go about your daily tasks. You will become entrained to the sight and sound of them and will notice quickly if anything has changed in their comings and goings.

Jacqueline has a tall log hive right out her kitchen window. I have colonies on my patios. We have never noticed our colonies being a problem with guests. The bees often join us, landing on our arms, or our garden tools as we work. As we unconsciously attune to the daily rhythms of bees near us, the bees attune to us and our activities, as well. In this way, our relationship with our bees deepens.

11

Do Not Disturb

A primary tenet in our style of keeping bees is what we call "very low intervention." That is, we keep out of our hives. This is not lazy beekeeping. Bees are extraordinarily complex beings in their hive culture, and our understanding of them is at a very basic, simple level.

When we dabble inside of a superorganism, we can't help but blunder. Such a complex creature is beyond our ken and requires that we approach it with reverence and restraint, and trust that 30 million years of evolution have provided bees with everything they need to adjust to any challenge Nature throws at them. Our inspections are, to the bees, chaotic intrusions into their day.

Throughout this book, we stress the importance of listening to our bees. Listening happens with the eyes, the ears, the nose, and the intuition. Bees will tell us that they are not interested in us poking our nose in their insides—which is a good way of describing the interior of a hive. If bees were pleased with us mucking about in their uterus (brood comb), pantry (honeycomb), and lungs (beeswax), we would not have to protect our bodies from their stings with hazmat-style bee suits.

Learning to leave the bees to themselves may be the hardest skill to acquire as a bee tender. But it is well worth the learning: First, do nothing. Then, do nothing a little longer.

Oftentimes, beekeepers inspect their hives to check on their progress. But in most cases, if a colony is truly failing, there will be little you can do to bring it back. If the colony fails, it was not you that failed them. Our approach with bees includes a willingness to let failing colonies perish. In these difficult times, bees are—like us—having to find new ways to deal with a shifting environment. Not all colonies will find the means to survive these times.

But keep in mind that bees have collective behavior patterns that naturally protect the entire colony from disease and parasites. Groups of bees within the colony can assess problems and create solutions on literally a moment-by-moment basis. For example, when sickly or aged bees die inside the hive, a group of undertaker bees will find them and carry them outside for disposal a good distance from the hive.

If another group of bees notices some of the brood have mites, they take on the self-developed hygienic behavior of opening the cell and removing the infected larvae, again taking them a far distance from the hive. Removing dead and diseased bee bodies keeps the hive clean and healthy.

Responding to diseases like chalkbrood that are caused by moisture, the colony can, as a group, increase the hive's internal temperature and overcome the pathogen. Each bee with her housemates works to do her share. And collectively, they accomplish large tasks.

Their healing capacity is also tied to their ability to seal off the hive with propolis, which lets the substance's curative properties bring about their healing. They are driven to be healthy and use their skills to do whatever is needed to survive.

How do they know who is healthy and who is not? Bees have a profoundly developed sense of smell. They know when flowers are at the peak of ripeness, when the nectar is ready, when an aging bee reaches her end-time. Guard bees know if an intruder is testing the defenses of the front door by sniffing her and discerning another queen's scent.

Nurse bees can smell when a larva is not perfect. They also can discern which queen larvae are the strongest and likely to be most fertile and they will focus their feeding on those queens, isolating and not giving as much care to the rest.

These tasks are all driven by bees' acutely developed olfactory sense, far beyond what we humans can even grasp as possible. Many of the animal kingdom are blessed with a phenomenal capacity to distinguish scent and their survival as predator or prey often depends on this sense. By meddling with the bees, we interfere with these natural senses and subsequently prevent their more protective behavior patterns.

Reflections from the Hive

Jacqueline

Last spring, I responded to a swarm call and found the bees clustered in a tall bush, all spread in a dense layer around and up the trunk about 5 feet high. This bush was a bristly clump of prickly needles, so I decided to wear my garden gloves. Oh dear. When I found them in the car, I realized I'd grabbed two left-handed gloves.

So, I said to my husband Joseph, "Do you have any gloves in the truck that you haven't pumped gas in?" He finds and hands me one of his right-handed gloves as he says, "These don't smell like gas."

So, I put on my left glove and Joseph's right glove, then reach my hands up into the bristly bush and start to lift a handful of bees between my two hands. As soon as I lift my hands, ten bees start stinging the right-handed glove. This is a clear and direct communication! The bees said, "Get that glove out of here!"

I briskly walked backward about 20 feet, a small cloud of sting-y bees nailing my right hand, and not one bee on my shirt, around my head, or near my left hand. Only about four stings got through the cotton part of the left glove, not so bad, and once I backed up far enough, I quickly took off my husband's right-handed glove.

That's when I asked Joseph again, "Are you sure you didn't pump gas in this glove?" and he answers, "Yeah, I have, but I couldn't smell any gas on it." He couldn't, but obviously the bees did.

I had heard somewhere that bees don't like the smell of gas. Based on my brief science experiment comparing un-gassy glove to gassy-glove, I now know that rumor to be fact.

From combined experience, we can tell you that although this hands-off approach can be very nerve-wracking to new bee tenders, after a few seasons the anxiety fades, as confidence in the bees and knowledge gained through hours of observation and study expand.

Fortunately, preservation-style beekeeping ensures that you will not be participating in activities that can harm your bees. By allowing them to manage their nest and interfering as little as possible, you grant them the greatest opportunity for success and good health. In fact, we usually find that bees the way we keep them just don't harbor disease, because they're able to go about their natural protective activities when disaster strikes.

If you are really tempted to dive into your colony, consider this: It will take the bees three to four days to put themselves back in order, rebuild their propolis seal, and reset the delicate temperature balance inside the hive. Should your intrusions cause the temperature inside the colony to drop, you put the brood at serious risk. Even a degree or two off the temperature required to raise the young will cause the baby bees to be born "less than." They will have less ability to think for themselves, to switch from one task to another, and their lives may be shortened, as well. Please. Let the bees be.

12

Acquiring Bees: Swarms and Bait Hives

There are several ways to acquire bees. Our preferred method is collecting bee swarms and hanging inexpensive bait hives in your yard and in places you believe bees may be visiting. Another good but more costly source of acquiring bees is nucleus (small) hives that local beekeepers offer for sale in the spring. The last and, sadly, most common method of acquiring bees is to purchase bee packages for sale.

We strongly recommend against this last method. Package bees are shipped in screened boxes grouped by the pound with an unrelated queen in her own small cage. Package bees may be a source of pathogens and diseases. They come from a different geographical area, thus the bees may not be prepared for your climate. They are all treated for mites with chemicals (a no-no in our practices, see page 151). And often, by the time they arrive in your hands, many are dead. If you can acquire no bees except for mail-order packages, we kindly suggest that you consider a different hobby. We are that adamant about the need to withdraw support for this inappropriate and deadly bee industry. Instead, we highly recommend these alternative methods:

SWARMS

As mentioned above, swarms are our favorite way to procure bees. Bees swarm to reproduce, and a healthy hive will produce a swarm or two or four in spring/summer of most years.

Swarms are an extraordinary event in the life of honeybees. We discuss them and their seasonal purpose more in-depth on page 29, but the best way of describing them is as the bee celebration of the year. A colony will take months of preparation before a swarm. In spring, a thriving hive acknowledges all the good work they have done and makes the momentous decision to reproduce itself by sending half the colony plus their queen out into the world to make a new colony. Think of it like an amoeba that grows so large that it splits in two.

Most likely, by the time you reach a swarm, it will have settled and appear as a shimmering dark cluster hanging from something. Before you got there, the swarming process was already well underway, with bees exiting their home hive and the queen summarily joining. Unlike the other bees, she hasn't flown since the prior year (or on her mating flight) so her stamina may not be as high as the foragers. She comes late in the swarm's departure for two reasons: (1) being a weaker flyer, she won't have to fly long before the swarm settles at a new location, and (2) she's well hidden in the chaos. To anyone observing, swarms look like pandemonium. The swarm's seemingly disordered commotion keeps the biggest bee hidden from bee predators, like birds, who'd love nothing more than to pluck and eat the big fat one.

The swarm unfolds from the whirlwind and heads to a nearby initial settling place. The bees will often not stay more than a half-hour in this first landing spot. The function of this short flight is to ensure the queen flies with them. On occasion, the queen decides to stay in the hive. In this case, the swarm will suddenly lift up and return to the hive. Once the bees know their queen is with them, they will often take to higher ground.

This initial swarm is called the Mother Swarm. Any additional swarms that come later (usually about 10 days after the Mother Swarm) are called Daughter Swarms, as they are led by the virgin queen daughters.

GATHERING SWARMS

There are books and videos that approach the gathering of a swarm of bees from a fully practical perspective, highlighting the tasks involved in collecting them and putting them into a new hive. Yet a swarm is a mystical experience as well, and we hope to give you a deeper sense of what working with a swarm of bees can be like. There is no other time when you will be as intimate with bees as when you are meeting them off their comb, and out of their protective nest.

Collecting a swarm is an experience of deep vulnerability, for you and for the bees, and you might notice how many times we use the word "slowly" in this chapter. This is not an experience to rush. Rarely does a person ever see a bee swarm in their lifetime, and far fewer will ever know what it is like to place bare hands on one.

We fully acknowledge that approaching a hanging ball of 20,000 bees for the first time can be a formidable challenge. We hope to make that prospect not only less intimidating to you, but actually one filled with joyful anticipation.

As to where to find swarms, we suggest you contact local bee clubs and see if they have a swarm list you can sign up for. Some swarm catchers advertise on local neighborhood email lists offering to rescue swarms. We actually get many calls a year this way.

If the swarm has made itself known, a beekeeper has two choices: let it go or collect it. Gathering a swarm and inviting them to live in one of our hives is one of the most sense-enhancing, gleeful tasks of a bee tender.

Working artfully with a swarm calls upon our ability to connect with the bees. Having said that, it's also probable that most anyone can capture a swarm. If you find a swarm in a tree, stick a box under the tree branch where the swarm is resting. Here is the easy way first: clip the branch and slowly place it in your swarm collection box. That is the gentlest way to gather the bees. If that is not possible, you can vigorously knock the branch to shake the swarm in. We have used an upside down basket as a swarm catcher, placing it near established colonies at shoulder height or higher. A swarm will sometimes make an initial stop and if you frequently check, you may find a whole swarm resting in a basket like this. Carry the box or basket to an empty prepared hive, dump them in and you're done. It seems plain and simple. But the process is really much more nuanced than that. For example, how do you "shake" a swarm of bees off a pole? Or a cyclone fence. Or the bars of a motorcycle?

In a typical swarm collection away from our own bee gardens, we ask if anyone knows if the bees have just landed or if they've been there awhile. Typically, swarms are stunningly gentle, as though they are in a trance of some kind. Even a swarm hanging for two days in good weather can be a gentle move. Approach with a warm heart: They understand the subtleties of our human states.

Some situations give us pause. Bees feel apprehension if they sense an incoming thunderstorm, if they are being moved toward dusk, or if the swarm has been out more than a day or two and the weather has been poor. Bees may be hungry or tired from hanging too long on a branch, or they may be cranky by nature. We often test the mood of the swarm by placing a cupped bare hand beneath the cluster and lifting just a bit. If the bees begin buzzing, moving swiftly, or stinging, we'll put on a bee veil.

Many swarms can be gathered by simply clipping the branch they are hanging from and placing it gently in a carrying box. If clipping is not possible, you can place your carrying box beneath the limb and give it a good, hard shake. Most of the bees will fall into the box. Shake the branch again, deliberately, until all the bees are off. Place a lid loosely atop the box, with one edge fully open. The bees will be in the air around the box for a bit. Watch to see if they begin moving toward the box and going in. Watch also for bees with their behinds in the air, fanning a scent to call the other bees "home."

Shaking a swarm into a box won't work when swarms land on surfaces that are convoluted, such as inside a pipe, spread over metal fencing, mixed into vine branches, or hanging under a table. Here it is useful to know how to move them with your hands. Mostly, moves like this will require you to be bare-handed. Gloves are just too clumsy for the delicate motions necessary for these situations.

Moving a swarm with bare hands requires patience and concentration. Be thoroughly familiar with swarms and capable of maintaining a sustained meditative state before you try this. We have a few guidelines for doing "scoop swarms" with bare hands, and being comfortable with bees comes first.

Start by using all of your senses and watching the slow movement of the colony. If the bees are too buzzy and restless, wait until they calm. Explain out loud to the bees what you plan to do. This gives you a moment of calm and stillness and broadcasts the energy of your intention to the bees.

Work with clean hands so you carry no scent the bees find agitating. Gently place your hand to the bottom of the swarm, cup the bottom of the swarm and let your fingers spread gently into the bees. When you go slowly, the bees will part as your hands enter and close behind you, folding you into the swarm, in the wonder of intimate engagement.

Leave your fingers loose and slowly move your hands left and right, not far enough to move any bees, but to let them know you can both move together. Slowly slip your fingers into the swarm to the second knuckles and gently bring your hands together, separating a section of the bees and lifting them upward and away from the cluster. Place the bees in a carrying box, or directly into a hive by gently fluttering your hands until the bees drop.

We keep a chunk of old comb in the bottom of our swarm carrying containers so the bees will have something familiar to congregate on.

A hive at the ready also offers bees an inviting darkness, and they may hurry off your hands into the interior. Often, all the bees will walk off your hands and onto the comb or into the hive. Then, return to the swarm and continue handful by handful.

At first many of the bees who enter the box or hive also slowly back out. This happens until we reach a critical mass of bees in the box, or until we have moved the queen in one of our handfuls. At that point, the process moves more swiftly; while you may rarely know when you have moved the queen, it's immediately evident to the bees and from then on, all the bees want to be in the box with her. The swarm will begin moving or fluttering.

Place your hive or carrying box right up close to the swarm, and watch them all march in! We like to leave the hive box at that location until late in the afternoon so all the scouts who have been out seeking a new home can join them. If that's not possible, close up your carrying box or hive making sure there is screening so the bees can breathe. Bees *must* have good ventilation or they can suffocate in a short time. And never put a swarm in the trunk of your car!

The scouts, upon finding the swarm gone, will find their way to their old home. An old farmer's tale says when the swarm flies to their new home, on a calm day the scouts can smell and follow the scent of the queen and thus rejoin the swarm later that day, but if you are carrying the hive away in a car, the scouts at least have the parent hive to return to.

Again: Please never put a hive of bees in the trunk of your car. The heat builds up faster than you would imagine, and it's very easy to suffocate them. They generate enormous heat and need good ventilation to cool themselves down. We generally put them in the back seat, tied in with a seat belt secured all the way around, and turn on the air conditioning. All our carry boxes and buckets are well ventilated with screening or small holes. If a few bees escape the box in your car, it's no problem. They will not sting you in the car, and when you get to where you are headed, you can catch them in a cup and release them next to their new home, once you have the main swarm settled.

HIVING A SWARM

When you bring a swarm home to your bee garden, there are two ways to hive them. You can gently shake-pour them into the top of your empty hive, or you can allow them to escort themselves in. Shaking them in has some elements of calamity involved. Although the bees pour out in a bundle like dark molasses, they may also take back into the sky in a plume, making you believe you've lost them. Be patient! They will settle slowly back down into the top of the open hive, at which point (maybe a half an hour or more), you can close the hive.

Another way to invite bees into your hive is to provide them an escort ramp. Let's assume that you have your empty hive off of the ground on a table or on cinder blocks. Place a wide plank of wood from the ground up to the entrance of your hive. Lay a bedsheet atop the plank.

Now, take your swarm of bees and pour or shake them from the carrying box onto the length of the plank. Many will take to the sky, but this is no problem. With your hands or a feather, move a few bunches of bees to the entrance of the hive, and with your feather, gently guide the swarm of bees up the ramp. They will want to head up, so this is quite easy. In no time at all, you will have a mass of bees happily marching up and into your hive.

While it seems counter-intuitive to dump your carefully gathered swarm onto the ground, this method of hiving actually works beautifully. As the bees march up the ramp and into the hive, the maidens all fan the Nasonov homecoming scent into the air as they walk. (The Nasonov gland at the very base of the bee's tail carries a scent something like lemony peanut butter. The bee pulls down the last scale on her abdomen to expose this gland. You will see it as a white dot near her behind.) Thus, placing your swarm on the ground gives them an invitation to choose the offered hive as their home, and they will rarely abscond from a hive they have chosen themselves. As an added bonus, while the hive marches up the ramp, you can sit and watch the bees and may even see the queen dashing up and into the hive.

If you have gathered your swarm by clipping the branch they clustered on, it will not work to place the branch on the sheet. The bees will stay on the branch because it carries the scent of their queen. You must shake them off, and hide the branch.

If the bees have landed on the ground (and this happens more than you might imagine), you can place a hive or collecting box on a stool, grab a plank and sheet, and encourage them to walk up and into the carrier. It is less effective to put the hive on the ground next to them and hope they will walk in. Something about walking upward is appealing to them.

In some instances, a swarm may have already chosen a different home somewhere else. In this case, they may come out and hang on the outside of your hive, or march up the ramp and settle on the hive instead of in the hive. We've worked with such swarms and sometimes just putting them back inside the hive a few times works. Sometimes, they just have other ideas and simply will not stay.

If you collect swarms all over your region, always have a well-equipped "swarm kit" in your car that includes a ladder or stepstool, clippers, rope, scissors, knife, feathers, a sheet, a plank, and a veil in case the bees are cranky.

Once you have hived a new swarm, leave them alone for a month. They may leave if you bother them too much during this developmental stage of colony creation. Sitting near them is fine, but avoid peeking in out of curiosity.

Gathering your first swarm is an adrenaline rush. All the swarms after that are bliss.

BAIT HIVES

There are many booklets and videos available on making and hanging bait hives. Our club fashions ours from plant pots made of recycled paper or hardened peat. These kinds of planters come in a variety of sizes. We like to use ones around 10 inches in diameter and 12 inches tall. But the dimensions are not too critical. Think small wastebasket size. To make a bait hive, we put two of these pots together like a clamshell, and connect them with zip ties. Inside, we place old comb, it will make it smell to the bees as though other bees have been there before.

Then, we spray a product called Swarm Commander onto the entrance of the bait hive. The essential oils in this product smell like queen pheromone, which is very intoxicating to bees. It is easy to find online, or in bee stores. We hang our bait hives with ropes or plant-hangers in trees or high up into patio rafters, and wait. Last spring, our club members captured nearly a dozen swarms in these simple, lightweight, inexpensive collection hives.

Once you hang bait hives, you must be sure you can check on them easily and regularly. Bees will begin constructing wax the moment they enter the bait hive. If you let them go for several days, then the bees are no longer in a swarm trance. They are in a protective mode, defending their new home. And they will be much crankier to move.

Also, once they are in "home mode" they will not want to move from the site. You will need to move them a good distance to a new hive (a couple of miles) or be sure to place many branches in front of whatever hive you have made for them, because they will orient themselves to wherever your bait hive was last hanging.

Sometimes, up to several hundred scout bees will explore your bait hive at one time, and you may think a bee swarm has moved in. Wait until the following morning, early, and watch that bait hive. Are bees coming out? At dusk, are there still bees entering the hive? If there are bees morning and evening, you most likely have a swarm inside.

You can also use a stethoscope to listen to the bait hive. If a swarm is inside, you'll hear it! If you notice the bees bringing pollen into the hive, this means they have already begun wax building and are crafting a new home. Shake or walk them into your hive soon!

LETTING BEES CHOOSE

Some people we know acquire bees in a different way. These bee tenders hang an assortment of bee-friendly hives like logs or Bee-Haven boxes, skeps, or modified Langs in their yards and on their land, and let swarming bees find them. It is the least intrusive way of bringing home bees, and is based on the notion that "If you build it, they will come."

Reflections from the Hive

Jacqueline

It is a sound like no other that comes from a hive—a buzz, a vibration, a sense of a roar, combined with tens of thousands of voices singing a high, sweet song. All that streams into your ears and your heart as thousands of bees spiral upward, covering the sky overhead in a quivering, dancing blanket of amber.

This is a time of racing heartbeats for the new bee tender. When the swarm first issues, there is nothing to be done but to breathe and watch. Allow the ecstasy of the bees to enfold you. Soon, there will be tasks to do, but please, do not bypass the opportunity to just be with the bees for a while.

Of all the events in my world, walking in a swarm is my absolute favorite. Excitement and joy are tangible, every bee flying in jubilant loops and curves. In my early years I wondered why swarms had so much commotion in them. Is there rhyme to their chaos? I've walked in many a swarm cloud, and never has a swarming maiden bee bumped me. Even at high speeds with 20 thousand bees, they swerve around me. Within bedlam, the bees maintain order. Each bee seems to have predictive skills that allow her to know the speed and trajectory of the other bees around her.

Sometimes, I need to shake the bees into a box. To anyone looking on, it may appear my method is no different than any old beekeeper shaking a swarm into a box, yet there is a fundamental distinction in the way these bees are handled.

I seek to deepen connection with them from the beginning, acting in good bee faith, unrushed and gentle, always respectful. I've worked with or seen videos of Heidi Herrmann, Michael Joshin Thiele, Ferry Schutzelaars, Debra Roberts, Gunther Hauk, Corwin Bell, and other magnificent beekeepers as they move bees, and I felt the holiness of their communion, how deeply connected they are in any small act involving bees.

I describe every step to the bees in great detail and ask for their permission and agreement. I listen for the tone of their buzz. Is it calm and harmonious, or ragged and multi-tonal? Do they seem interested in continuing their meditation, or are they concerned about me? Are they smooth-mantled, or rough-edged and fussing? When I gently wave my hand through them, do they ignore me, or do they seem worried? Each response tells me if I have permission to move with them. When I find all in agreement, I move forward.

Not all swarms lend themselves to re-homing. If I feel them too distracted or notice the swarm ramping up their activity, I take that as a sign that they are readying for departure, and I let them go. I once spent three hours alone in a tree with a gorgeous glittering swarm whose presence filled me with joy. Despite being completely prepared to catch them, I never felt agreement from them to do so. I felt it was their intention to move to the land out beyond our farm. Their presence entered and filled my heart so deeply, even today, I still feel blessed by that connection, just as I felt honored to witness their departure.

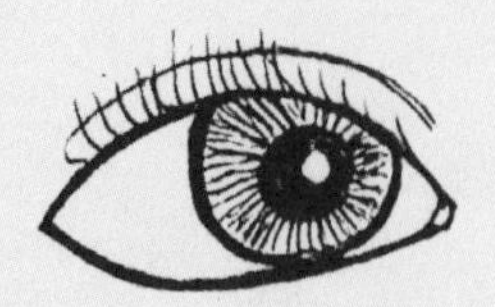
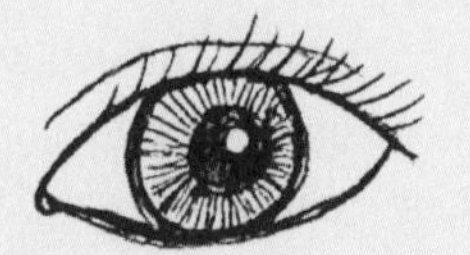

13

Bee Language

Slow down! A busy mind and quick, rushing movements will not serve you well in your bee garden. I've never learned anything while I was in a hurry.

The most obvious aspect of bee language is the language of the body: bees are always moving, touching, doing. And a huge part of bee language is indeed their body language and the sound of the vibration they make in and about the hive.

Learning to speak "bee" comes with time. As a new bee tender, there is so much of a mechanical nature that needs to be sorted out—what kind of hive, where to put it, what bees, when to start?—it is hard to find the time and the ears to hear this new language. But if you sit with your bees and spend dedicated time in observation, it is a language you can't help but learn.

Jacqueline and I speak emphatically of the need to observe, but many people have little idea of what close observation looks like. It is a skill you can come to master with your bees, and it will serve you for the rest of your life as well.

As you do this, it will become near impossible not to develop deep affection for the bees. Seeing the complexity of their lives, their genuine and tireless work on behalf of the whole, we can begin to feel the depth of their bee-ing.

Bees are supreme teachers of measured, calm behavior. Being with bees in bee time has made us far more patient, kind, and peaceful people. While I do not have the intuitive gifts of Jacqueline, by any stretch of the imagination, I have been able to find my own unique way into bee language—and you can, too.

The kind of observation Jacqueline recommends is, perhaps not coincidentally, the same kind of mindset one works to achieve in meditation: a non-conceptual mind that sees with no preconceived story attached.

As soon as you try this kind of observation, you will realize, as I did, how every single thing that comes into your thinking has a story attached to it, an attitude, a judgment. It is our human condition to endeavor to make sense of what we are experiencing. Our self-made stories about things are our way of bypassing the unsettling, startling sense of groundlessness that can gobble you up once you set your stories aside. But this is where I believe bees are asking to meet with us: in this landscape of utter "don't know," where we can meet the mystery in each other.

Bee language works both ways. We must listen, and find a way to answer back. And in the process of learning how to listen and how to respond, the bees will repay you by making you a better version of yourself.

Humankind has the ability to communicate on much deeper levels than spoken language. It is a skill we can reignite and remember as we seek to be in a deeper relationship with our bees. Below, we share our own particular means for talking with bees.

THE IMPORTANCE OF OBSERVATION

We are enthusiastic advocates of the "thousand hours" club. Jacqueline in particular is blessed with everlasting curiosity, and she works hard to partner her inquisitiveness with a good dose of open-mindedness to avoid being presumptuous. I always want to know how, why, where, and for what purpose things happen, especially with my dear bees.

Being open-minded is a crucial part of observation. If I can let myself bask in "not-knowing" for a while and not name a behavior, I sometimes notice distinctions that I would have missed if I only "saw" bees bringing in pollen. As soon as I name what I think I see, I stop looking at that behavior with an open mind. For this reason, I try to see each action, each situation, as if this is the first time I'm looking at it. When I open up my senses, I become privy to a fuller perception. Today I may notice the smell at the entrance has a minty sweetness to it, which will send me wandering over to look at the bee activity in the mint beds.

If it's a hot day and I smell a resinous cottonwood scent wafting about, I may wonder if the bees are blending and moving propolis. That makes me wonder if they are concerned about comb stability and have shored up the upper comb attachments and edges of cells to prevent comb from falling in the heat. If that's the situation, this would not be a good time to open a hive.

I spend days watching bees do all kinds of daily tasks, and I don't find one minute of it mundane or boring. Spending a few hours a day observing bees may seem over the top, but it doesn't stop there. Sometimes I record what I'm looking at, come into the house, and watch that same video a dozen times more.

This is where asking myself a handful of questions is key. I enjoy this questioning immensely. Are there helpers at the ready? Are they cleaning the front entrance? Are the guard bees having a casual day? If they are on alert, can I figure out what the threat is? Could it be a yellow jacket buzzing overhead, or is it an incoming rainstorm that has them on edge? When this bee touches antennae with that bee, what transpires? Where does each bee go after that communication? Do either of them then pass that info on to another bee? What does that third bee do?

From these questions, I can identify patterns. Is it a pattern I am familiar with? (Lots of dandelion pollen coming in?) Is there a hive behavior I can associate that with? (Lots of pollen is a good sign they are rearing plenty of brood, many babies are hatching!)

At first, you'll see bees landing and going inside. You might notice they are bringing in different colors of pollen. This tells us which blooming flowers they prefer this time of day or this time of year. Watch how a pollen-laden bee lands. If the pollen bag is heavy, that bee may have a hard time maneuvering and will swing wildly in the air before landing in a tumble on a bee below.

When I first started watching bees many years ago, I watched simply because I was fascinated. I saw all kinds of stories unfold on the entrance board. A bee wobbles and lands on another bee. At first, I thought it random, and then a hundred wobbling bees later, I noticed a dozen details that explained why that happened a dozen different ways. When I was a novice, I didn't see the details, but the more I watched, the more explanations made themselves clear. Some examples:

- A flying bee helicoptered in, holding her balance as well as she could, considering she had a dense and heavy load of pollen on each of her thighs. Ah, yes, the day's task is dealing with a higher than usual pollen harvest.
- The bees on the landing board were hyper-excited. Many of them watch carefully to determine if the flying bee is friend or foe. Oh! These watching bees are guard bees, ready to defend the front door.
- A wobbly bee flies in that is much bigger and less coordinated. Now I see it; this is a drone!
- The flying bee has a different balance point coming in. Perhaps she has nectar, which is weighted toward the front of her body. Now I see she is balancing two heavy loads.
- Too much traffic coming in. Why is the air so full of bees? Many

seem to be hovering in front but not landing. Could this be the time of late morning when the 16-day-old new foragers first venture outside? I notice they are orienting themselves to the front door so they have a visual of the entrance.

- Are they coming in close to landing, then lifting to a different area or angle, then making another approach? Hmm . . . and does this seem to happen each day at about the same time?
- How come so many bees are going inside in a rush, but hardly any are coming out, even though it's midday? Did the barometric pressure just change? Is there a storm coming?

These are the things I observe as I watch. I notice what calls my eyes to see. I hope you will find this process rewarding as I do. With persistent study, you will begin to make sense of what used to look like a hundred buzzy bees in chaos. Suddenly you see the same hundred bees, but now there may be 16 different things going on that explain themselves to you.

Reflections from the Hive

Jacqueline

I've been tending bees and teaching classes for beekeepers for many years. I started before the internet was popular, so online groups and YouTube videos weren't yet available like they are today. I struggled to find the kind of education I wanted. I read everything I could get my hands on and spoke to most anyone who had bee experience, which gave me a bit of knowledge, but my desire was to know how the bees wanted to live, and finding that out was a challenge.

Early on I set a chair next to the front entrance of my hives where I spent hours watching and trying to interpret what their actions said about their tasks, mood, intention, emotional state, health, and temperament. From this experience I realized that spending "a thousand hours" in the presence of bees was the best way to learn about bees. I don't mean a thousand hours going inside the hive and doing things to the bees; I mean a thousand hours of hands-off observation.

I was in a land where I didn't speak the language. Bees went in and out, and that was about all I saw. When something unusual happened, like the first waggle dance on the entrance board, I wasn't very good at interpreting. I guessed that information was being passed along, but what, where, and why?

Normally the bees ignored me. They seemed to think of me much like a tree, landing on my sleeve every once in a while, but no real interaction. Then one day they seemed churlish, and I got the feeling they didn't want me there. But how did I know that? Not taking the hint, I sat still in my place to figure out how and why I would think that. What was different today than the day before?

On that day I noticed a few differences. Every few minutes a bee flew straight at me and bumped the front of my protective hat's screen. I thought those bumps were accidental, but I wondered if they might be intentional. Then I noticed their sound seemed a tad higher-pitched, and where they normally walked with their legs comfortably wide and bent, on this day their legs seemed straighter, like they were walking taller.

Over time I came to realize that these cues are part of their language of high alert. Indeed, they were talking to me, telling me that my presence that day was a disruption and they'd prefer that I go elsewhere. My bee language skills weren't developed enough to know why they pushed me away that day, but I got the message and moved off.

How handy for me that I understood. If I'd had a task to do in the hive that day and done it, I certainly would have picked up a few stings, caused a bit of an in-hive uproar, and wondered what I did to inflame them. I would have thought I'd done something wrong inside the hive, but they were giving me fair warning outside the hive and telling me to leave them alone that day.

STINGS HAPPEN

We pick up a piece of comb and surprise a bee resting underneath it. We walk barefoot in the clover field and step on a bee. We get too close to the hive entrance and a bee takes exception to the presumed intrusion. We are collecting a swarm (swarming bees rarely sting) and inadvertently squeeze a bee. Inevitably, both the bee and the bee tenders are surprised!

We get a few stings a year, but it seems fewer each year. That is probably because we've learned to pay more attention to each bee around us, trying our best to bring them peace rather than discord.

We choose to work with no bee gear as often as possible. We don't like the feeling we bring into the bee garden when we armor ourselves against them. When you are suited up, you can bungle a lot, and the bees cannot clearly tell you, "Stop, not now!" In my bee garb of tank top, shorts and bare feet, I bring myself to the bees vulnerable, trusting, and very focused on their language.

I honor their venom and the fact that they can cause me harm if they choose. It keeps me humble, meek, and utterly present. In this way, I tell them in my own language, "I'm here not to force my dominion on you, but to work side by side with you. I come to you with no protection."

However, we do recommend that beginner beekeepers don some kind of bee gear to protect their less experienced hands. A hat and bee veil, bee jacket or suit, and gloves, as un-clumsy as you can find. While we usually work around bees without protection, there are even times we don the hat and veil. Comfort comes with time and experience. Always be gentle, and don't hurt a bee when you are around them.

READING THE BEES

With rare exception, what I bring to the hive is what I reap. If I am preoccupied with other thoughts, I am not paying attention and can easily miss signals that should have told me to back up, withdraw, slow down, or simply go away. Bees give the signals; it's up to me to read them.

Before I visit the bees, I assess my own emotional situation. Have I been grumpy all morning? Do I feel frustrated, anxious, bereft? Am I stressed for time and hurrying? Have I recently argued, or am I worried? Am I happy? Happy is the one that works best, anything on the plus side of neutral. If you are leaning toward sub-neutral and bringing an emotional burden into the bee yard, it might be a better idea to do something else

But in my flawed humanness, I sometimes carry emotional baggage into the bee yard. When I do, the bees have two responses: calling me in closer, or suggesting I come back another time. A wise bee tender knows the difference!

Reflections from the Hive

Susan

Last year, I tended 20 swarms of bees in my yard. Tending swarms is happy, time-consuming work with a lot of associated tasks: watch the bees issue out of the hive, track them with your eyes (or your feet if they've headed off down the street), wait for them to settle, determine how best to gather them. Will you need a ladder? Clippers? A friend to help? Ropes to pull down a high branch? A saw? Where is the hive you will put them into?

On this particular day, I watched a swarm arise like a departing, vibrating soul out of Charity hive. She rose up swirling in the air, and I could feel my mind start to click off the tasks before me. But in that instant, another thought arose strong and swift, blotting out my usual leap into frenzied doing: Put the tasks aside. Bee with us.

So, I stood up and walked into the midst of the swarm that was now directly overhead. I raised my arms up and began singing to them, a Lakota healing song pouring from my throat. They answered with their swarm song of deep, ecstatic vibration and began sinking slowly downward until they were all swirling around my body.

Time went away. Thoughts went away. I went away. It was just sound, and singing, and vibrating. I felt the pulse of 10,000 wings fluttering around my face and arms, and watched the hypnotizing blur of their bodies moving in syncopation and chaos at the same time.

Enfolded in bees, I was in a place of utter vastness, of no boundaries at all, yet it was not a scary place because the bees were holding me together in some comforting, sweet way.

Suddenly off to my left, my husband called out to me to ask me some question about . . . something. My first reaction was annoyance. My concentration broke, my arms started to fall to my sides, and a host of bees buzzed into my face and rapidly stung me. Startled, I backed away, hands to my face, flooded with a sense of shame: I had broken a trust.

I have never had bees behave that way toward me in the midst of a swarm. Mostly, swarming bees ignore me and allow me to be in their midst as they would allow a flagpole or a fencepost. But not this time. They had invited me deeper, and I had followed, and then I had broken the moment with my negative reaction. And they wanted me to be certain that I knew what had happened.

I have been a serious meditator for years and the bees have been my most profound teachers in this. The attitudes that most beautifully serve a human life—kindness, respect, compassion, devotion to the greater good—are attitudes you must develop to be in safe company with bees. Bee language is the language of love, and love betters all things.

Despite the condition I bring to them, the bees may respond with forgiveness and call me in. I notice this by identifying that the bees ignore me and go about their ways, gently wafting through the air, perhaps landing on me as if I am a tree branch. Of course, they know I am there, but they find no reason to pay me much attention. I enter preoccupied and, without me even noticing, I slip into bee time as clarity and presence return to me.

Other times, the bees determine in the first few seconds that they don't want me there. They show with their behavior that they prefer I leave the sanctity of the bee yard and return when I am less preoccupied—which I do.

If defensive behavior or little stings occur, let them know that you understand this gentle direction by leaving the area. This response tells them you understand their message. It's the start of a two-way bee communication.

The way bees posture themselves on the landing board can tell you a lot about their emotional tone or the task they are doing. Bees with their tails tipped up and pointing out to the fields are misting droplets from their Nasonov gland, scenting the air to help foraging bees identify the way home. This is a happy task for everyone.

Slightly different is when the hind end is up, up, up all the way to the tip. That is a threat to sting. The bee is saying, "Here is the part I want you to notice, my stinger!" What is the proper response? Back up immediately.

Did a bee bump you? Bees have no sense of scale that tells them that you, human, are 10,000 times their size. A bee bumps to say, "Back off now!" My experience is that you have about a half second to respond and briskly walk about 20 feet away. Pause too long and the next bump comes stinger-side first.

Gareth John is a UK bee tender with lovely bee language skills. He says bees rearing up on their back four legs with the two front legs raised means, "Don't come in," while bees walking toward the entrance with their front end down in a posture of submission (as seen in dogs) means, "Please, may I come in?"

This is where observing your bees, with no task agenda, is richly rewarding. Watch long enough and you'll see all kinds of fascinating body language and behavior. I am finding myself hesitant right here to explain all the different bee postures and signals because, rather than having you memorize "this means that," it's far more amazing if you watch long enough that you start to understand movement-phrases bees employ.

You may see a foraging maiden scurrying in figure-eights to share the location of nectar and pollen. Yes, that is fascinating, and now watch how each individual bee responds to her waggle dance. How is that tidbit of knowledge conveyed to the rest of the foragers?

Pick a bee and follow her (or him) as she weaves her way through the throng. Who does she respond to and how? Do this with 20 individual bees and then do another 20.

I have no rules for spending time with bees, but I can suggest guidelines that will make the time more enjoyable for all.

Settle yourself before you enter the bee yard. Have you just had an argument, read something disturbing, or feel distracted? This is not the time for a bee visit.

Pay attention. What is the bees' temperament as you approach? Are they quietly going about their tasks? Is the sound calm and easygoing? Is the sound robust and busy, as when they are fully deployed to the fields and each bee comes back fully loaded with nectar or pollen?

When bees are tremendously busy and on task, this is not the time to butt in for a cup of tea. Let them work. Are there agitated guard bees defending the front door? Don't interfere. Do they seem to be frantic about something? Can you tell what it is? Don't increase their stress. If there is a fuss happening, be the calm one and send them love and kindness.

Reflections from the Hive

Jacqueline

I have a video I took one afternoon of 10 minutes of easy-going activity at the hive entrance, an equal number of bees casually flying out to the fields and coming back loaded with pollen and nectar. Then in a scant moment, I felt the air change. I wouldn't say I am always that perceptive, but somehow this subtle information made it to my awareness.

Looking up, I saw in the distance that a rain cloud was headed our way, still about 10 minutes out. The barometric pressure must have fallen suddenly, because the returning bees immediately doubled the speed at which they were flying home, and not one bee exited from inside the hive to head out to the fields. A single raindrop on a bee's back can knock her to the ground, so these field bees were flying at hyper speed to get back home before the rains came.

Take a few moments to feel the bees' presence. Does your intuition tell you anything? Can you align that with any visible behavior? Do you feel an invitation? This, too, is part of developing the bee language, where asking and responding take place.

Spend time in their presence. Are you capable of quieting your own thoughts? This open yet quiet mind you bring is the canvas upon which the bees may lay thoughts to you. That is how it often happens for me, when my own mind is quiet. I may have a thought pop in about pollen, or I may feel joy rising in me.

14

Water for Bees

You might not think of your beehive itself as being a source of water for your bees, but in a good bee-centric nest, bees become gifted water engineers. Bee colonies in natural nest configurations work hard to fully line their living space with propolis, which is naturally waterproof. Propolis can then be used to channel water that collects on the moist sidewalls of a colony into small collection pools.

Bees generate a great deal of water in their nests from respiration, nest humidity, nectar evaporation, and hive cooling. Bees are able to take all this water, and literally turn it into medicine by channeling the water along the propolis rivulets they have crafted. As the water moves over the propolis, it gathers potent antibiotic, antifungal properties. Today, you can buy "smart" or "infused" water at the grocery store. Bees have been crafting smart water for millions of years. Propolis water is medicine made by bees, for bees.

In winter, when it is dangerously cold for bees to fly, the bees can use their own water sources inside the hive. You might think that all this moisture would be deadly to bees in the hive, and much conventional bee literature talks about the dangers of cold condensation dripping on bees over winter, and freezing the colony to death, or mold building up on cold honeycombs.

However, new research is painting a fresh picture of bees and water in the hive. Where bees are living in nests in which they can maintain stable hive temperatures (heavily insulated hives, see page 86), and the ceiling of the hive remains dry (like with a quilt box or skep, or a hollow log), the bees create a dome of warmth in the upper half of their nest. This stable warmth, and the medicinal qualities of propolis, enable the bees to create sterile air inside the nest, defeating pathogens before they can even arise. In this sterile environment, no molds grow, and the combs can remain pure for a long time. In such a healthy environment, bees utilize to their full benefit all the water inside their nests.

But our bees will need more water than the water they create in their hive. Honeybees do not like to travel far to collect water, and will search out the closest consistent water source to their nest.

WATER SOURCES FOR BEES

Before you ever bring bees home, create a watering station for your bees that will be close to their hive. Bees need water to drink and to keep the moisture in the hive air at the right humidity.

Why bother building a watering station? If you don't provide water for them, your bees will find somewhere else to source it. If they find your neighbor's swimming pool, you'll soon have cranky anti-bee neighbors. Also, your bees may get into polluted or poisoned water after someone washes their car, drains antifreeze, or leaves tainted puddles around.

It's so easy to give them the water they need, and at the same time make your yard more beautiful. Here are some water station ideas we use, and what it cost us to make them.

Plain Old BirdBath

Twenty dollars at the hardware store. Really simple. We add rocks, seashells, moss bundles, and a few wood branches so that bees have a safe place to land and drink. The only caveat is that you must remember to fill it. Bees will find another watering area that is more consistent if you allow yours to dry out now and then.

Stock Tank

Sixty to a hundred dollars at a farm store, plus a $35 fountain that keeps the water moving so we don't breed mosquitoes. We've added pond lilies and irises and local duckweed for landing zones. If you don't want to bother with a pump, add a few inexpensive goldfish from the pet store and they will eat mosquito larvae. A nice feature about stock tanks is that you never have to remember to fill them daily.

Ponds and Fountains

We've found old fountain ponds at garage sales or on neighborhood sale sites. Sometimes, you can find fancy double-tiered things for surprising little money.

Bees prefer to keep their feet out of the water and like to stick their tongues in the tiny cracks and fissures in the concrete at the water's edge. That way they don't risk falling in. In the concrete pond, the bees can stay a few inches from the pond edge where the water is drawn by capillary action up into the cracks. No need to get their feet wet.

Concrete Pavers

This is really simple and you can often find used pavers for free. Stack concrete pavers a few layers tall, so they make a cube. Lay down a sheet of pond liner large enough to overlap the edge, and add another course of pavers on top so the liner edge doesn't show. Fill with water and plants that like wet feet. Suggestions are mint, calla and canna lilies, asparagus fern, native duckweed, and watercress.

Keep a bucket of seaweed almost covered with water. The seaweed releases salts and trace minerals from the ocean that are beneficial to bees. You can also use shallow dishes filled with small pebbles so the bees can put their tongues down into the little crevices while standing safely on the pebbles.

CHOOSE A GOOD SITE

Choose a location for your watering station that's out of your yard's traffic flow, or you will be walking through the flight path of bees and other thirsty insects all day. Light shade or at least afternoon shade is best. In shade, the water won't evaporate as quickly. Once it's established as a watering site, the bees and many other creatures will come to count on it, so please remember to keep it filled.

SAFETY

Fill the watering stations with platforms of gravel, stones, and moss. That gives the bees something to land on, and then they won't fall in and drown. They like to land and walk to the edge of the water to drink. Moss is particularly good because they can stand on top of it and drop their long tongues down into the water.

We've seen bee tenders use corks, marbles, quartz rocks, and charcoal for bees to stand on. Floating pieces of wood, porous lava rocks, and plants all work beautifully. Remember to keep your water station chemical free. Don't use aluminum pails or painted containers that might leach chemicals into the water that can harm bees.

BEAUTY

We encourage you to make your water station beautiful. Our culture has lost touch with the ancient knowledge that beauty is healing. Our efforts to create beauty in our bee gardens reflect our good intentions to the bees and make us more respectful bee-tenders.

Decorate your station with quartz crystals, statues, flowering plants, native moss, and water mint. Be creative! There is a calm, delighting peace in making things pretty. Give this gift of beauty to yourself and your bees.

15

Plant Like Crazy

When people tell us they want to help bees, we tell them to plant lots of bee-friendly flowers. Tending bees in hives is not for everyone. It takes a lot of learning and dedication. But anyone and everyone can plant flowers. Even if all you have is an apartment balcony, you can grow pots of flowering herbs.

Bees inspire one to garden, and gardening is good for bees, wonderful for the environment, and healing and nourishing for the gardener. Gardening for bees is like any other kind of gardening, but with a few considerations.

GARDENING FOR BEES

When planting for bees and other pollinators, it makes sense to know what is already growing in abundance in your neighborhood. Do your neighbors have flowering trees or berry bushes? Are there huge, abandoned lots full of blackberries? During what seasons is your neighborhood in full bloom? When does the flowering decrease? In our neck of the woods—the Pacific Northwest—late summer and autumn are when our bees are most in need of blooming forage. We have many native trees and bushes that bloom early in the spring and summer, but then the flowering decreases abruptly, and the bees start becoming more desperate for forage.

This is the time of the year we see bees at hummingbird feeders, on fallen fruit, and on chicken feed. So, we let our neighbors and their willow, filbert, and fruit trees feed our bees through the spring. We plant heavily in fall-blooming plants like autumn sedum, buckwheat, asters, goldenrod, pineapple sage, and borage.

Learn by trial-and-error which forage plants are particularly happy to grow and prosper in your own yard, and grow a lot of them: Borage, oregano, phacelia; and large stands of goldenrod, fennel, and hemp agrimony are easy to grow in almost any garden.

Focus first on native plants, which will provide the best forage for our native bees. For ease of gardening, plant perennial bushes and flowers.

With these hardworking plants, all you need to do is to prune them as needed and welcome their green shoots each spring.

As important as what to plant, is how to plant: honeybees prefer a large patch of a single kind of flower to a yard studded with one each of every pollinator plant under the sun. This is because honeybees gather one type of flower on each nectar or pollen flight. A single plant does not catch their eye, as there is not enough food there to make it truly worthwhile to them.

A plot roughly the size of a single or double bed makes all bees very happy. If you fill such an area with any bee-loving plant, you will be treated to the sounds of happy humming as long as the blooms last.

Through our studies in permaculture, we've learned that nature abhors bare dirt. She will quickly plant something to protect and to feed the soil. Mostly, we call what she plants weeds; and so, we follow her lead in our gardens and allow flowering weeds, which are some of the most nutritious forage for bees.

Weeds move into disturbed or depleted ground to root deep and cleanse the soils. The die-back of weeds in the fall nourishes the soils below. These wild plants move quickly to fill in areas between planting beds: dandelion, nipplewort, bittercress, mallow, crane's bill, wild geranium, purslane, and chickweed—all are cherished by our native bees as well as honeybees.

We have always disliked the look of "maintenance-free" yards, all planted with junipers, shrubs, and wide expanses of weed-sprayed lawn. In these all-too-common (and all-too-boring) yards, each plant is isolated from her neighbors by yards of barren bark dust. Science now tells us that roots like to shake hands, sharing information and nutrients under the ground. We imagine these orphaned plants must be lonely. Each must stand alone, eat alone, and bear the winter winds alone.

Out in the woods, we see the plants cozying up to each other, jewelweed peeking through the nettles, creeping jenny winding affectionately around a skunk cabbage. Imagine whispered green dialogues amongst all the leaves, and bees sharing in the intimate conversations.

Plant like crazy. Plant what grows easily for you. Plant in bed-sized plots. Leave no dirt naked. And learn to welcome the flowering weeds.

IS THERE ENOUGH FOOD FOR ALL?

Honeybees are not native to North America, and the only legal designation they are granted in this country is as livestock. Lovers of native bees and other pollinators are concerned that honeybees are damaging populations of our native bees.

It is true that a production beekeeper with several hundred or even thousands of hives can upturn the ecosystem wherever they settle. Often backyard keepers find their hives robbed out by huge populations of hungry production bees fed constantly on sugar syrup.

But if you are keeping your bees in small, well-insulated hives, they will not require near the honey stores that production colonies must gather to survive winter. By making the most appropriate honeybee home, you are also helping the native bees to thrive, and there will be enough food for all.

We recommend keeping a watchful eye on your forage plants. Do you see lots of bumblebees, and even the tiny native bees the size of a small ant with wings? Can you identify native bees and pollinating wasps? You'll find good identification guides in our resource section.

If you are seeing few native pollinators, it may be time to allow more flowering weeds. Let a section of garden or driveway go to weed, and watch who shows up there!

Plant some of your yard in mixed-flower beds that honeybees do not heavily forage. Native bees don't have the directive to gather only one source of pollen or nectar during a foraging flight, and are happy to forage in beds with a thick mix of plants.

Be aware of the balance of honeybees and other pollinators in your yard. Keep fewer hives if you notice the number of other pollinators dropping.

THRIVING SOIL, MESSY GARDENS

If you are gardening on ground that has been abused or neglected, you may be surprised to find that bees are not visiting all the new plants you've provided for them. One reason honeybees are such perfect pollinators is that they only gather from the healthiest of plants. If you are working to reclaim your ground, it might take a few years for pollinators to really start flocking to your plants.

Weeds help restore soil. Dandelions pull toxins up from the ground, dock roots break up hard soil. One technique for building up healthy soil is called Chop and Drop: whenever you cut back plants or dead flowerheads, chop up your clippings and leave them in the garden. This is compost in the making.

Never waste fallen leaves. They are garden gold. Spread them on planting beds, over compost piles, and under berry bushes. This is how nature makes her own soil, and you can help by not being so quick to throw away clipped grass, plant clippings, and leaves.

True gardeners for bees will be known for their slightly wild yards. Weeds are welcomed; grass is full of clover, heals-all, dandelions, and wild geraniums. Come autumn, the spent brown plant stalks remain standing and dried leaves are gathered and dumped onto planting beds. In these hollow dried plant stalks and beneath the shelter of the leaf piles, pollinators are overwintering. Winter birds flock onto the dried seed heads. Spiders and bumblebee queens are hibernating. Soil is being quietly constructed from the ground up.

If you think about it, winter is when creatures are most vulnerable. By offering nesting places, winter food, and shelter, your garden will be brimming with life and renewed energy by spring.

Reflections from the Hive

Susan

I hold to a belief that no plant or animal finds you by accident. If such a one shows up in your life, then that relationship matters. When we first came to MillHaven, I kept my eye out for a nice little spearmint plant for me and the bees. I started several that I purchased from a variety of stores. Mints are nothing if not hardy. Who can kill a mint? Me. I killed them all. No matter where I placed them and no matter the care I offered, they quickly turned into shriveled sticks.

But suddenly one spring, a luxurious mint took hold among the irises, and by next spring, she had crept beneath the gravel path and popped up next to a very old azalea. So much lusher than any mint plant I ever brought home from a nursery, this plant squeezed out the irises and jumped over to claim several other old garden beds. And I let her go as wild as she liked.

All summer, I cut this mint and put great handfuls into a gallon jar of sun tea water. She is my summer elixir. When she blooms, every pollinator in our neighborhood is there to celebrate among her spear-like flower heads. When I brush past her on my way to the garden shed, she sends her fresh, tangy perfume to follow my footsteps.

I offer thanks to whatever mysterious presence knew we needed this green friend here at MillHaven. Perhaps it was the spirit of the bees themselves. Surely the bees know how much I love this garden, this home. The plants must know it, too, by my caressing touch and my earnest care of them. Every gardener, and perhaps every insect, knows there is much magic in well-loved ground.

16

Offering Food and Gathering Honey

Bees eat honey. They craft this precious food from flower nectars mixed with enzymes, and evaporate 70 percent of the water from the nectars before they determine the honey has ripened, at which time they put a wax seal over it to indicate the honey is complete.

Once the honey is ripe, it has a shelf life of thousands of years. Naturally antibiotic, antiseptic, antifungal, and antimicrobial, honey is a wonderful wound dressing, useful in preserving mummies, and a powerful medicine when taken in small amounts daily. It is the elixir that makes it possible for bees to survive through winter when flowers are no longer blooming.

Bees have a fragile digestive system, and have evolved to subsist on one of the most easily absorbed foods in nature: flower nectar. Pollen is also collected by bees, but they do not consume it in its fresh state.

Inside the hive, pollen is mixed with enzymes, a bit of nectar, and bee saliva and head-pressed into comb cells where it naturally ferments by a lactic acid process much the same way as kraut. More than 8,000 different microorganisms have been identified in this pollen ferment called "Bee Bread," making it a perfect probiotic and high-protein food for adult and baby bees. This pollen has a chewy, tart taste, and is a much more digestible form of pollen for humans, as well. Bees prefer fresh pollen, less than 72 hours old, and studies indicate that bees who eat fresh pollen are healthier.

In commercial honey-collection operations, most of the honey the bees create is taken, and the bees are fed back various mixtures of sugar and water as syrup or a candy-like fondant. The reasons for

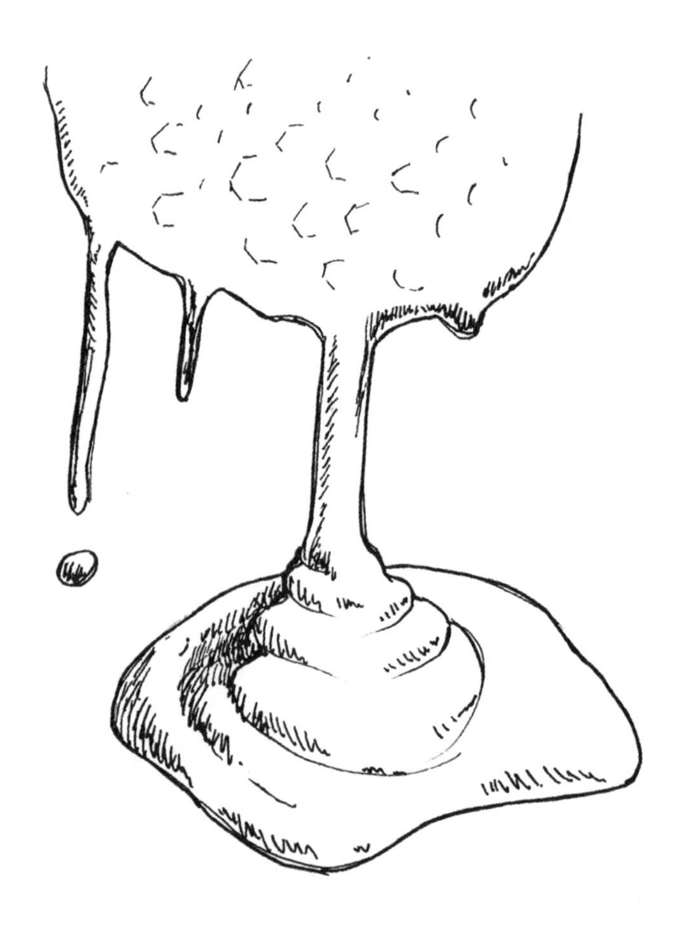

this are entirely money-driven, as honey is much more expensive than sugar so it pays the beekeeper to take the honey and feed sugar back to the bees.

And because this method of sugar feeding is so money-driven, you will find a lot written supporting the feeding of sugar. For-profit beekeepers simply cannot give it up. But sugar is hard on the guts of bees. The pH is entirely different from honey, and it has none of the nutrition of honey.

To put bees in the risky position of feeding them sub-par food during their most challenging, life-threatening season of the year seems like madness to us, and we avoid it. We want our bees to come out of winter as healthy as possible, and they will not do so on sugar.

Ideally, because you have created a stress-free lifestyle for your bees, they have had the time and the energy to collect all the honey they need for winter. Bees must make important calculations going into the winter that consider the number of bees in the colony, the honey stores collected, and the bees' best estimation of how long those honey stores must last.

We often have late and very wet springs here in the Northwest. Our gardens may be in full, glorious bloom, yet rain and cold keep the bees inside. When the sun arrives between showers, the nectar and pollen sources are compromised for hours or even days, and so April is often known in bee yards as the Starving Moon.

New beekeepers often don't know enough about noting the weather in relation to bee activity, and are stunned when a hive that has lasted all winter suddenly perishes when the fruit trees bloom but the rains are heavy.

Though I didn't follow it at first, Jacqueline and I now practice a rule around feeding our bees: we avoid it. We let the bees keep their honey, taking only small amounts for our families, or gathering the honey from perished hives. But in situations of very odd weather, a very late swarm, or a very poor forage year, we will feed our bees.

Watch your bees closely in spring and autumn. If you suspect it's been a poor forage year for them, and if your hives feel feather-light if you raise a corner, feeding is a good option.

WHAT TO FEED

When we do choose to feed, we feed our bees honey. As a new beekeeper, you will most likely not have much spare honey around, and in these cases, we recommend that you purchase honey from someone who sells it "raw and unfiltered." Honey that is filtered is heated, and heating destroys all of the vital enzymes in the honey, turning it into something much more like sugar.

Yes, honey is expensive, but if you craft your own hives, and get your bees free from swarms you collect or bait hives you hang in your neighborhood, you can put some of your saved bee-dollars into some good honey. You will read warnings in conventional beekeeping books about feeding your bees honey, because it may contain deadly spores from bee diseases. We find that these bee diseases come from large commercial operations, not from small backyard enterprises. We have never had any ill effects from honey we have purchased at local farmers' markets.

So don't be tempted to buy honey from the grocery store. There are books and documentaries out there that will tell you all the ways that honey has been tainted or diluted in the commercial distribution process. Find a farmers' market and get to know a local beekeeper.

You'll learn that feeding your bees can be a bit messy, both for you and for the bees. It is important to place straw, twigs, or stones in bowls of open honey, or the bees will fall in and drown. Honey still in the comb is best, but you'll rarely have it on hand for the exact days you need it. Ask me how I know this.

You can keep clean old beeswax out of perished hives, and pour honey on the combs in a bowl. Or you can use frozen honey balls if you have them.

You may be tempted to slightly dilute the honey, especially if it has crystallized and hardened, but don't. Adding water encourages fermentation. You want your bees to have honey, not mead! Also available as artificial feed for bees are pollen patties. But bees who are fed this fake food source do not do well and have shortened lifespans. Stick to honey, full strength, in emergencies only, and you will know all you need to know about feeding bees.

In the unfortunate case that you find you have starving bees and no way of procuring honey for them, the Natural Beekeeping Trust offers a recipe that can be used if your bees are truly in dire straits (see our Resources on page 180).

HOW TO FEED, WHERE TO FEED

Where to place the bowls or balls of honey is easy if you have eco-floors in your hives (see page 88). The bowls go down on the eco-floor. Overall, it is best to feed bees inside their hives. There are times of the year when it is safe to put out open bowls of honey for multiple colonies, but you don't want to try this until you've had a few years with your bees.

Open feeding can set the bees into a feeding frenzy. The bees get so excited and aggressive over the food, that they will often rush off to the doorways of the other colonies to see if they can break in and get more. Robbing is a scary business and hungry bees can steal all the honey from a weak hive in short order. Open feeding will also attract a lot of yellow jackets, who will be more than happy to eat your bees along with their honey.

Another very simple and effective in-hive feeder we have used is a Mason jar attached to the outside of the hive. A jar ring is screwed to the side of the hive, a small hole is drilled into the hive/Eco-floor to allow bees access to the jar, and the jar is screwed into the ring. When you need to feed, unscrew the jar and place the honey or honeycombs into the jar for the bees.

We especially like this feeding method for two reasons: you don't have to enter the hive in any fashion, and you can easily see when the bees need more food.

GATHERING HONEY FROM A PERISHED HIVE

If you have been keeping bees for a few years, you will have honey of your own saved that you can feed. We rarely gather honey from an active colony. But if you keep bees, you will find that hive losses are sadly common. This is when we gather honey. We collect from perished hives. Some folks will put a small box atop a log hive or a skep and gather a bit of honey in that way. In any case, once you have the honey-heavy combs in your hands, you will need to process them.

Here is a simple way to process raw honey easily, and without the need for special tools or strainers:

- Put all the combs in a large pot, and mash them vigorously with your hands or a potato masher.
- Pour this honey-mash into a large strainer set over a collecting bowl.
- Place the honey bowl somewhere warm in your kitchen and allow the honey to drain.
- When the honey is drained, then gather all the remaining sticky comb into tennis-sized balls and freeze this for bee food. The frozen balls can be placed in a bowl, or offered in a ball jar on the side of the hive for feeding.

We don't promote using honey as a sweetener for baking or even in hot drinks. Cooking and heating honey will instantly destroy all the enzymes that make honey the powerful medicine that it is.

Use it this sacred way, as medicine. Honey makes an excellent covering for wounds and especially punctures and burns. Take a spoonful for stomach upsets or coughs. A spoonful at night is said to help with sleep. And our experience has shown us that local raw honey is a powerful ally in the treatment of hay fever and allergies.

17

Treating Bees with Chemicals—A No-No

As you've surely noticed by now, few of the alternative hives we advocate for are easily accessible for bee "management," and we've already spoken about our preference for hands-off, eyes-on bee tending. When we discuss these topics in new bee forums, these are the comments we get:

"If you don't treat, your bees will all be dead in three years."
"You wouldn't keep medicine from your children or pets, why do you refuse to offer it to your bees?"
"You are cruel and misguided!"
"I'm so glad I don't live near you and your poor, neglected bees!"
"If you don't inspect, how can you check for illnesses?"
"Your neglected bees are making my hives sick"
"How do you get honey?"

Our short answer to all of the above is this: Our management style IS our treatment:

- We tend bees in small hives.
- We encourage swarming.
- We don't take our colonies' honey and replace it with sugar.
- We keep our bees warm and dry and up high.
- We provide eco-floors and insulation.
- We keep out of the hives.
- We provide as much organic forage as our gardens can grow.
- We trust natural selection to decide which hives are fit and which are not.

All of these actions result in less overall stress for the bees. Stress is a killer. We want our bees to be free of as many of the human causes of stress that our innovations can eliminate.

BEE AFFLICTIONS AND THE MIGHTY 600

In the early 2000s, a mite named *Varroa destructor* made it to the shores of America and Europe from Asia. Asian honeybees (*Apis cerana*) have been well-adapted to the mites forever, but Varroa was a game-changer for European honeybees in North America (*Apis mellifera*) who had never developed a relationship with this fast-breeding, body-fluid-sucking creature.

With the coming of this little red mite, bees in the wild and in apiaries alike died by the millions. Scrambling for solutions to this destructive pest, which not only weakened bees but infected them with a myriad of deadly viruses, beekeepers turned to the chemical industry, which has been supplying them with various insecticides for nearly two decades.

But bees are insects too, and it is not easy to find the appropriate level of insecticide that eradicates Varroa mites but does not kill "too many" bees. The treatments are harsh, and the results spotty. The poisons never kill all the mites, leaving alive the ones who are resistant to the miticides. Thus, poison product after poison product has been tossed aside as the mites develop immunity to them. The bee industry has, in its misguided efforts, created super mites who are still feeding off of our bees.

There are a host of beekeepers working to develop "mite resistant" queen bees, but this involves mechanical insemination of queens instead of allowing them to naturally reproduce. When bees are kept naturally, as soon as your hive swarms, you have a new queen who will mate with wild, not-necessarily-mite-resistant drones.

Meanwhile, what about the wild bees who have no one to douse them with insecticides? Estimates are that with the coming of Varroa, more than 90 percent of wild hives perished. The loss was swift and catastrophic as migratory beekeepers transferred Varroa all across our nation in only one year.

But while Varroa is still destroying managed hives all over the world, a different story is emerging from the wild hives. Research by Professor Tom Seeley undertaken on the Arnot forest near his Ithaca, New York, home shows that the wild bee populations have bounced back to pre-Varroa levels.

The big takeaway from Seeley's research is that these bees, left in the hands of natural selection, showed more than 600 changes in their DNA over a 30-year period from when Seeley first took samples from these hives years ago. Six hundred adaptations to manage the Varroa challenge! Seeley stresses the wonder of this dynamic process that occurs between bee and mite when we let natural selection rule. There are no domesticated bees that have shown anything near this kind of adaptation.

Recently in a bee chat room, we heard a long-time keeper say, "the solution for Varroa will come from a laboratory." Even with all the evidence piling up that it won't, most beekeepers today use insecticides in their hives and verbally attack those who choose to let nature be the physician.

But by allowing our bees to work out their Varroa issues themselves, we are encouraging the best, heartiest genetics in our bee gardens. Bees who have not evolved the tools to manage the mite perish quickly. Those with mite-fighting skills survive and their swarms then populate our new hives.

These strong bees have developed tools like grooming each other for mites, biting and chewing the legs off of mites, pulling mite-infected larvae from the sealed brood cells where the mites breed, and swarming often. Each year, scientists are reporting new mite-managing strategies found in unbothered hives. Recently, they've discovered that bees are now uncapping and recapping larvae. The larvae are unaffected by this, but it disrupts the mites' egg laying.

Swarming creates a natural break in the breeding cycle as new comb is built for egg laying, and as the mother hive grows a new queen. Varroa can only breed inside the bee cells, so when no bee eggs are being laid, no Varroa eggs are, either.

And when Varroa can't breed, they go sterile. It takes a month for them to rekindle their breeding capacity, so you can see just how big a benefit swarming confers!

Other bee pests you should know about include American Foulbrood (AFB), and Nosema. AFB is a fatal bacterial disease of honeybee brood caused by the spore-forming bacterium *Paenibacillus larvae*. It is not a stress-related disease and can infect the strongest to the weakest colony in an apiary. Infected brood usually die at the prepupal or pupal stage.

Heavy infections can affect most of the brood, severely weakening the colony and eventually killing it. The disease cannot be cured, meaning that destruction of infected colonies and hives or irradiation of infected material is the only way to manage AFB.

AFB is characterized by a foul smell coming from the combs, where the dead larvae are decaying. Sunken, perforated caps on the surface of the brood cell are seen, and if you stick a toothpick into the cell, the dead larva will pull out rope-like and stringy.

Foulbrood is often called "The Beekeepers Disease" because it spreads easily hive-to-hive in close quarters. It is rarely found in wild hives where bees set up nests a good distance from each other. We don't know what other qualities may be present in a wild hive that act as deterrents to foulbrood.

As you get more familiar with your own bees, you will learn that bees smell good. Everything about them smells of flowers, honey, and tree resins. The aroma wafting from a hive is positively delicious. If it begins to smell otherwise, you likely have a problem.

Nosema is another illness that smells bad, like old, sweaty gym socks. Nosemosis, or Nosema disease, is caused by two species of microsporidian parasites (a type of spore-forming fungus) called *Nosema apis* and *Nosema ceranae*. *N. apis* is thought to have originated on European honeybees, while *N. ceranae* is thought to have evolved as a pest of Asian honeybees (*Apis cerana*) and has only started to affect the European honey bees relatively recently. *N. ceranae* appears to be more damaging than *N. apis*, affecting more cells in the bee's mid-gut and killing infected bees faster than *N. apis*.

Both of these Nosema infections cause dysentery-like symptoms in bees. It is most noticeable in spring, when bees have been confined to the nest for weeks and months on end. As they begin joining the outside world again, it takes them time to clear their guts from the stagnation of winter.

Bees will not defecate inside the hive, but will hold their waste for weeks until a tiny breath of winter warmth allows them to fly out, eliminate, and fly back to the hive before their flight muscles stiffen. In early spring, there is a lot of pooping going on.

Often in spring, you will find streaks of bee feces on the face and entry of the hive, as bees rush outside to the "outhouse" but don't quite make it. It normally clears up in a week or two.

With AFB and Nosema or dysentery—in fact with nearly all of the bee ailments, both common and uncommon—the best means of prevention is to keep your hives healthy and well-fed. Nosema worsens with sugar water feeding. Also, bees crammed closely in managed yards spread these illnesses quickly among hives.

If you are following our guidelines for keeping bees, you should rarely if ever encounter bee ailments, outside of Varroa mites, which are in all hives, all the time.

What does it look like to be a new preservation beekeeper and choosing to go chemical-free? We will be very honest: at first, it looks scary. You are new to bees. Everything you see them do is new to you, and there is so much to learn! It takes time to learn what is normal behavior and development in a new hive and what seems "off." In the beginning years, you will often lose a lot of hives through problems not of your own making.

Jacqueline lost all of her hives during her first four winters with bees. I lost all my hives for several years running. Come spring, we acquired swarms and started all over again. We see that there is a sort of alchemy that happens as a few years pass: we bee tenders learn more and become more relaxed with our bees, tweaking our methods and hives, planting more flowers. Meanwhile, we bring new genetics into our bee yards each year.

Eventually, a few of those lines of bees take hold in our yard. It all flows together from many directions: right hive, right place, right bees, right tender, big love. The stars and the planets align. Suddenly, your own bees are swarming and surviving season to season.

You stop worrying about mite life cycles. And life is good. You just need to be patient enough to weather the first years of tending bees, when you seem to live in a constant state of anxiety and hair pulling, and every dead bee causes you upset. Trust us, it gets better. Much better. Now, let's switch gears and talk about the poisons we put in our yards, the poisons which are now saturating our air, soil, water, and plants.

HEALTHY GARDENS, HEALTHY LIFE

In addition to avoiding chemicals in our hives, we are doggedly organic in our gardens as well. Over time, we pray more people understand that gardening with poisons is a losing situation. Follow this chain of misery: bees bring back pollen and nectar laced with trace pesticides (either grown into the plant or sprayed on) to their hives. The combs quickly absorb the pollutants from the stored foods, because that is how beeswax works. Within a year or two (or less, if you live in a heavily polluted area) the comb itself is deadly.

We believe that growing organic results in healthy organisms and healthy food for our bees. Here's the formula in a nutshell for how a healthy organism grows:

The seed is planted, it draws nutrients from the soil, blossoms, fruits and sets healthy seed that starts the whole process over again. When a disease or pest threatens it, the plant is strong enough to resist infection or thwart the pest.

Here are a few ways a weak system works:

Weak Seed: A seed is planted, but the seed has weak genetics that compromise the plant right from the outset.

Poor Nutrition: A healthy seed is planted in weak soil, which means the plant doesn't get enough nourishment and it develops into a weak plant.

Poor Conditions: A plant starts out healthy, but is unable to deal with withering heat, freezing cold, too much rain, or lack of water.

Poison: A plant starts out with a healthy seed and good nutrition, but gets exposed to poisonous chemicals that damage the plant's ability to respond to its environment.

People often overlook the role Nature plays in each of these situations. Nature wants living beings to survive to the next generation. When a plant or creature becomes compromised, Nature wisely sorts them out of the gene pool.

The gene pool is Nature's gift to the future. Ideally, only the healthiest beings should be part of the gene pool. Pests, bacteria, viruses, and fungi are guardians of the gene pool. In Nature's design, these mechanisms are called in to dispose of the unhealthy thing, thus preventing the seeds or spawn from becoming part of the gene pool. The strong and healthy ones get to reproduce, and if the environment is inviting, that process carries on into the future.

For billions of years, Nature has followed this path. Healthy life begets more healthy life. Weakness is culled out, generation by generation. In this way, we can allow our gardens to function naturally much as our bees do. And we find that a healthy, more natural garden also results in healthier bees.

POISONS OR LIFE

Jacqueline and I have both lost colonies to poisons. Even small amounts of toxic chemicals, whether used to treat mites or brought in on poisoned pollen, can have a cumulative negative effect on a hive. Bees carry everything they touch home to the hive on the bottoms of their tiny feet.

Bees tainted by garden chemicals will arrive home and ask for assistance getting noxious sprays off them. The very nature of bees is to help each other, especially when a bee is in distress. So, the home bees start grooming the sticky foragers, trying to get the poison off. The bees' communal activities put each bee in contact with the next, easily spreading the spray around the hive. What's more, the soft warm wax inside a hive acts like a sponge, absorbing and holding the toxic chemicals. In a few hours the entire colony will be contaminated.

We ask you to consider the following, and we hope it makes you as queasy as it makes us. Some bees have developed a new behavior in recent years, never reported in the past. Bees in certain hives have begun entombing their pollen behind propolis caps. Once sealed, the cells are never opened again, because the bee bread itself is so tainted with poisons that it has become deadly to the bees.

As chemicals persist in a hive environment and continually expose the colony and new bees to poisons, it further weakens their immunity. Any toxic chemicals, even from "safe" in-hive treatments, can cause ongoing harm.

A recent study examined beehives from 23 states and two Canadian provinces. They found 121 different insecticides in the bees, wax, pollen, and hives. Not a single one was free of chemical exposure. Sixty percent of the pollen and wax samples had at least one systemic insecticide. One of these systemics, neonicotinoids, is lethal to bees, moths, and beneficial soil insects, yet they are still in use in the USA, though quite a few European countries have wisely taken this and other bee-harming insecticides off the market.

Read a label the next time you're in a store that stocks lawn and yard care products and try to make sense of it. Unless you're a chemist, there's no way any of us could be expected to understand the implications of these poisons on a bee's delicate system, or the rest of Nature's wee creatures. The tiniest amounts, measured in parts per million, are devastating in ways we don't understand.

One study found that toxic effects of a certain chemical didn't seem to damage bees who had contact with it. Rather, the effects showed up generations later through damage to the bee's DNA. When studies are done on products like this, real time exposure and damage would seem to say these chemicals are harmless, yet the damage is buried in bees not yet born.

Also, the EPA is not required to study synergistic responses between chemicals. This means that we citizens have no idea of the chemical cocktail we put on our lawns (emergent herbicides, grub killers, weed sprays) are actually creating new cancers beneath our feet.

And speaking of feet, do you feel safe letting your children and pets walk barefoot in city parks? We don't. The research into how the EPA and USDA conduct studies for new agricultural poison sprays and soaks is woefully depressing. Much is missed in our particular version of "the scientific method."

As bee tenders and as earth stewards, we must become more responsible about our use of toxic chemicals—for the bees' sake as well as our own. Smart solution: go organic and avoid chemicals altogether. The more you can build the health of your trees and gardens, the better off bees are. If your trees have an accessible bounty of nutrients and minerals, your orchard and the bees who pollinate them will thrive.

Reflections from the Hive

Jacqueline

Like many folks, I used to think poisonous chemicals surely were illegal and that the EPA and FDA wouldn't allow toxic chemicals into our hands. I was surprised to learn otherwise, and now I spend time educating others about that fact.

The governing agencies responsible for approving pesticides and herbicides know certain chemicals are dangerous, but producers can get them approved if the label gives specific directions that explain how to use that poison correctly. That puts the responsibility for proper chemical use squarely on the shoulders of the consumer, not the company.

Orchard tree sprays, especially ones that have bee-harming chemicals, are designed to be sprayed at very specific times and in certain ways so they reduce the bees' exposure. Yes, the sprays do harm bees, but if used properly, the exposure may be limited. Unfortunately, studies show that most (one study says 96 percent) consumers don't read directions.

While writing this, I looked up the directions for a commonly-used toxic fruit tree spray. The directions are eight pages long, in small print, and the average person likely wouldn't make it to the end of the directions. But there on the very last page you'll find this warning:

"BEE CAUTION: This product is highly toxic to bees exposed to direct treatment on blooming crops or weeds. Do not apply this product or allow it to drift to blooming crops or weeds while bees are actively visiting the treatment area."

Most people would never guess that the innocuous-looking spray bottle with the pretty flowers on the label can kill entire bee colonies.

Most people have never seen a pesticide kill, but I assure you, once you see one, you will do anything you can to prevent it from happening again. Here's what it looked like when this happened on my farm.

A few years ago, on a sunny Sunday morning, a few forager bees came home with poison on them. By late afternoon bees hundreds of bees had dragged themselves out of the hive, falling onto the ground, writhing in death throes. Within 24 hours, all were dead. A pesticide kill is horribly, tortuously, agonizingly slow to a single bee, yet ridiculously fast in killing 50,000 bees.

I have a good guess as to what happened. Somewhere within 2 miles, a fruit tree went into blossom. Bees are attracted to flowers, not buds or leaves, so if you must spray, the timing is crucial. If a tree is in bloom, assume bees are already on it. If even only a few bees, less than 1 percent, visit the sprayed tree, they can carry home enough poison to kill the whole hive.

Likely, the local tree owner near my hive didn't get around

to spraying at the proper time, before the bloom. A week later the buds turned into blossoms, sending out the signal for bees to approach and gather nectar and pollen. Bees came from far afield to drink the nectar, spread the tree's pollen, and fertilize the orchard.

The tree owner probably walked out Saturday morning and noticed that the trees were in full bloom and remembered he forgot to spray. The spray directions caution against spraying when the tree is in bloom (flowers are a bee attractant), on a sunny day (when bees are out), or when there's a breeze (wind carries poison). Still he figured "better late than never" and pulled out the sprayer. He dowsed the tree as the bees busied themselves pollinating it. That's the recipe for a bee disaster.

The result: my bees, and probably dozens of other nearby bee colonies, including native pollinators like mason bees, butterflies, bumblebees, and possibly even birds who eat poisoned insects, die.

Beloved scientist Jane Goodall wonders why we ever thought poisoning our food was a good idea. We wonder the same. And the solution is right in our hands: put down the cans and buckets and squirt bottles of poisons and trust nature. She's been doing a fine job of championing life for billions of years.

18

When a Queen Dies

The bees' relationship with their queen is beyond our ken. All of the maiden bees in the hive forego their own reproductive directive to support the queen in hers. A hive can survive if it loses maidens and drones, but if it cannot grow a new queen, the colony is as good as dead.

The queen is everything to her colony: mother, creator, and cohesion builder. It is her scent wafting like a lemon-buttery elixir that holds the hive in trance, spurring them on to their work of bringing abundance to the world at large, and their colony in particular.

It is not a stretch to say that colonies with strong, healthy queens are happy hives. You can hear it in their song, and see it in the steady, efficient movement of the bees on their comb and in the foraging fields. If their mother is well, then all is well.

WHEN A QUEEN NEEDS REPLACING

In production hives, queens are killed by the beekeeper at about one year old and replaced with a brand-new queen. This annual queen replacement happens year in, year out, ruled by the calendar. If for any reason the beekeeper thinks that the young queen is inferior in any way, the beekeeper kills that new queen and replaces her with a third queen. The beekeeper determines the timing and reasoning, not the queen or the colony.

If all your observations were done in production hives, you would feel confident that you understood when and why to remove an old queen and replace her with a new queen. In this conventional scenario, there would never be a situation where two queens lived in that colony. You would never see the way bees handle queen succession on their own.

If you don't manipulate the conditions or inhibit their natural behavior, the colony will reward you with all manner of heretofore unseen situations.

How do bees decide when a queen needs replacing? After five to seven productive years, the colony notices their queen's fragrant scent becoming weaker. Sadly, because of pesticide exposure, drone bees are not as fertile as they used to be—even if the queen mates with many, many drones, she may still not be able to fill herself with a full complement of semen. In this case, she can find herself unable to lay any more eggs before the summer season is even over. Her diminishing scent tells them the queen's fertility is slowing down. Or perhaps the queen has become ill or injured in her movements through the hive. In these moments the bees know it is time to replace her with a new queen, the process called supersedure.

During supersedure, nurse bees search for the right egg to make into their new queen. Not just any egg will do. They seek an egg from what Cornell professor Tom Seeley calls the "royal caste," a special and rare egg not found in great numbers in the hive. While any fertil-

ized egg can become a queen if fed with only Royal Jelly, some eggs are just more suited to royalty than others, and the bees recognize this.

But humans have no capacity for discerning which of the eggs are royal and which are common. In the bee lab, the next egg in line becomes a queen. That human-chosen queen lacks the ineffable magnificence of royalty known to the bees. The manufactured queen is missing something the bees deem as necessary. We don't know what queenly skill or adaptive trait got left behind in the choosing.

Through supersedure, natural colonies choose the correct princess egg and they plan the correct timing. The new queen is prepared to take the old queen's place before the old queen's fertility runs out, giving the colony a smooth transition. Once the young queen is hatched and mated, once the nurse bees confirm by her scent that she is decidedly fertile and her eggs are healthy, the old queen's caregivers cease feeding the old queen royal jelly, further reducing her hormonal vigor.

The old queen is still full of life, but no longer responsible for reproduction. The house bees feed her honey just as they do the rest of the hive bees and she, at last, is permitted to roam at will, free of any further queen responsibilities, until she dies a natural death. As preservation beekeepers, we have been treated to the sight of multiple queens in a colony.

WHEN BEES ARE UNSUCCESSFUL MAKING A NEW QUEEN

Beekeepers are grossly mistaken when they say that you can save a failing hive by giving them a new queen. By installing a new queen of human choosing, the beekeeper renders the colony and her history essentially dead. If the new bees choose to accept her, the new queen will begin laying eggs quickly, and none of them will be related to any of the bees in the hive. Their genetic story will be entirely new.

Everything about the reproductive drive of honeybees is meant to ensure the advance of *their* ancestral line forward either by swarming, by drone, or by supersedure. If a beekeeper adds a new queen to a hive, the existing colony continues to exist only as servants to a whole new family of unrelated sisters.

Whatever genetic gifts the maidens and drones possess are lost when the queen is replaced by an outside, unrelated other. And so, we take the act of requeening a hive (or choosing not to) very seriously.

There are a few reasons why bees might be unsuccessful in making or keeping a new queen. Queens can be lost when flying out for their week of initial mating flights. It only takes one hungry bird. Sometimes robber bees or yellow jackets rampaging in a hive can kill a queen. Sometimes, if you mess about in a colony at certain times of their development when the colony is, for lack of a better term, less cohesive, the bees may turn on their new queen and kill her, equating her with the sudden calamity. Mostly, queen problems in our bee gardens have been with virgin queens either not breeding well or not making it home alive.

Neither one of us have had any success with introducing a new queen to a colony that has been unable to craft one on their own. We've both tried, and followed all requeening instructions to the letter, but our bees say "no," so we've stopped trying to convince them. Yet others we know have had success in adding new queens to a failing, queen-less hive.

Here are some of the things we've observed with queen-less colonies in our own bee gardens. First, as we've already noted, we have both noticed that often hives will contain two queens. Perhaps this is a new development with bees, but we've read that a number of hives create a new queen late in the summer and take her and her mother through the winter perhaps to ensure that they have at least one functioning queen come spring.

We have both had numerous experiences with watching a colony shrink down to a small handful of bees with their queen. The small cluster of bees tend their queen and care for her, but don't seem to be able to coax out of her some queen-making eggs.

We believe this may well be the result of the amount of background pesticide pollution in our environment these days. Queen bees used to be the longest living bee in a hive, enjoying six to eight years of productive laying and living. These days, we're told that queens reared in queen-rearing operations are rarely lasting a year.

Many of the pesticides and other chemical treatments in our environment have been proven to disrupt the ability of the queen to produce, to find her way home to the colony after her mating flights, to remain healthy and robust, to live long among her sisters. These chemicals also disrupt the ability of bees to absorb nutrients.

Drone bees are implicated in this reproductive challenge as well, since those same chemicals also suppress their sperm production. A queen may mate with as many as 15 drones on her mating flights, and still run out of viable sperm in her first year of egg laying. We wonder if these are the queens nestled in with a tiny cluster of sisters in a failing hive: confused and ill, quickly reduced to infertility, exhausted in a mere matter of months.

Queens are fed only royal jelly produced in the head glands of maiden bees. Being traditionally the longest living bee in the colony, you would be correct in assuming that the queen is subject to more pesticide contact in the food she is fed by pesticide-compromised maidens, and in her constant contact with pesticide-laden honeycombs.

When we talk about healthy hives and healthy bees these days, it's sad to think there may be very few of these left on earth. Because we humans tend to seriously downplay the effects of pesticides in our food, water, and air, we can be oblivious to the horrendous load of poisons every living thing is now exposed to each day (to read more about this, see page 151).

If you read bee forums about hives that have failed, nearly everyone responding to the post will say two things: "What were your mite counts?" and "It was Varroa." We don't buy into this knee-jerk response to hive losses. Maybe Varroa was the straw that broke the bee's back, but certainly we need to count pesticides, poor nutrition, and poor beekeeping practices as the root source of most of our bee losses.

And so it is with queens. Singular disasters like queen loss are often addressed with singular solutions like adding a new queen. Yet the survival challenges queen bees face are multiple and cascading, from poorly insulated hives to poisoned food, poisoned combs, weak drones, and weakened daughters.

In a failing colony, eventually one or more of the maidens may begin laying eggs. All the offspring will be drones, as all the eggs of the maidens are infertile. But by sending a huge convoy of drones out into the world (where they will hopefully mate with new queens) some of the bees' ancestral line is preserved and all is not lost. Once the maidens begin laying drones, they will in no circumstance accept any new queen you may offer.

In presenting a deeper, broader context to the subject of failed queens, we hope to provide you with a clearer lens with which to view any queen issues your colonies may present to you. Whether to requeen or not is a choice. And an important one.

With a queen-less hive that is unable to craft a new queen on its own, you actually have several options:

- You can try to add a new queen
- You can choose to merge this hive with another
- You may add a swarm of bees to the colony
- Or you can let the hive perish.

All of these choices take some planning and work. If one fails, you may try another. They can all lead to dead ends, or they can all succeed wonderfully. You never can predict with bees.

Reflections from the Hive

Susan

We won't advise you which of the four aforementioned paths to take with a queenless hive. But we will strongly advise this: sit down by your bees and ask them what they would like, and how you can help. Sitting with these questions, leaving yourself open to images, ideas, or memories will open doors in the mind and allow right actions to emerge.

I sat in this way with such a colony last summer and clearly felt an intuitive sense to just leave them be. Other times, we've each sensed a colony's desire to merge with another hive, and we will do that when we feel strongly directed by the bees.

When we speak of being directed by the bees, you may wonder if we are simply responding to our own projections. Who knows? Yet we all know when something just feels "right." Not "good" or "needed" or "wanted" but simply "right."

We believe it is wise to behave as bees do and to follow what we clearly sense is right. If nothing seems "right," then do nothing. Do nothing until you get that "aha" moment, and move forward on that. If such a feeling never comes, assume that doing nothing is your best choice—your right choice.

Epilogue: Returning to the Bien

Susan: In the small confines of my city yard, bees have been leading me by my passion for them into new realms of awareness. First, they made me keenly aware of their love for the flowers, and which flowers they loved most.

I came to know these plants because of the bees. I learned their blooming and dying seasons, in which parts of the yard they flourished best, and I happy-danced each spring at their return.

Each plant was no longer just a stalk with leaf and flower, but an old friend, returning spring after spring for more adventures. Watching the bees on my flowers, I next came to notice all the different pollinators who also appreciated my flowers: Ants, wasps, beetles, spiders (hiding out for an insect meal), praying mantids, hummingbirds.

Planting on my hands and knees for my bees, trowel in hand, I have become friends with the soil and all the creatures who live there. A worm slides smoothly across my fingers. A beetle scurries away from my knee. A bright red dragonfly comes by to see if I've unearthed anything tasty.

This is the bien. A German word, the *bien* (pronounced "bean") considers not just the hive, but the whole area the hive serves as the "bien." While the word "hive" leaves me out as the bees' tender, the "bien" brings me into this village.

In following my heart deep into the heart of the bees, I am suddenly in community with thousands of small, important, mysterious beings who have such an important role to play in the health of this Earth. I am part of the bien.

I believe bees invite us to this: to return to the community of the world from where we originated. The bees reveal to me there was a time when all living things worked for the good of the whole, and this is something I choose to believe, and must believe. The bees teach me hope that we are, were, and will be again "one"—the holy and eternal one from which all our better angels arise.

Now, I garden for all the six-leggeds, the winged ones, the creepy crawlies. Now, I make piles of spent leaves and old twigs for winter beds for all those little beings who call MillHaven home.

Now, I place water for all. I place tiny altars and create places of beauty with stones, statues, and carvings, all about the yard to remind me MillHaven is a sacred place inhabited by sacred beings, myself included.

At the top of our yard stands a large pollinator condominium, a gift to all the insect beings who visit and live with us. The condo is tall with a peaked roof and shelves, and holds dozens of cans, pots, and wooden rounds all filled with hollow bamboo rounds, dried valerian and foxglove stalks, paper straws, and even old birdhouses where red-rumped bumbles like to nest.

A bench sits beside the condo, and I sit there often in the summer watching all the many societies of insects who use the hollow tubes for nesting. At the condo, there are no arguments among wasps, bees, flies, and beetles. All are welcome. Mason bees, wool carder bees, sweat bees, leaf-cutting bees, grass-carrying wasps, so many more I can't identify, and even opportunistic spiders all call the condo home.

Lulled by the singing of 10,000 insects, I look up in the heat of summer and see tiny, translucent orbs floating in the clear sky and I whisper to myself, "the very air is alive." Of course, it always has been. I just never saw it until the bees showed it to me.

In whatever years I have left, may I work to make this world as rich, tender, and sweet as the bees have made mine. May we all work as bees do, in service to love. And may we each find our way back to the whole and wholesome community of the bien.

Jacqueline: After my first book, *Song of Increase,* was published, I had the merry delight of perusing reader's book reviews.

Immediately I saw something I hadn't anticipated.

When I wrote the book, I wanted to share my understanding of how best to serve the bees. That was, and continues to be, my intention. But I noticed that many, many of the reviews and comments had a common thread. I will share four so you get to know the experience they each had and why I am including them.

One reader's comments are a good example:

"Jacqueline's words always resonate and reassure. Watching her is like sitting near my hives. One cannot help but breathe deeply and inhale her love for her bees that is so immersive; it completely surrounds me in such a way that I feel I am inside it and part of that luscious goodness within."

Another reader described how the book "reassures me in spite of the chaos in my head and in the world around me. Nature knows what she's doing and it's going to be okay."

Another . . . "I so appreciate that [the author] allowed this deep connection with the bees to have voice and expression. Through this example we each receive inspiration for our own inherent connections with this wisdom."

Another . . . "The connections between the cellular and the cosmos are presented with simplicity and sweetness."

Each of these people cites a common theme: We want to know we are a part of Nature, a piece of the whole, and that our bee tending demonstrates the awareness that we are One. We want to step across the bounds of species with open minds, and hearts unfettered by pretending a steep hierarchy separates us.

I have sat with bees for more than a thousand hours. Their presence stirs in me a deep and true reverence. I feel blessed by the company of one little bee standing on my finger, sending out her long tongue to taste my skin scent, lifting a mid-leg to neaten her hind-leg. Many, many times we have looked eye to eye, she with ten thousand hexagonal lenses, me with two.

When I was in my early 20s, I wondered about life on other planets. Would other beings ever visit ours? Could we humans step from behind our fear and welcome someone from another galaxy? Could we bring peace to one another?

I didn't want to leave that to chance, so I took it upon myself to start practicing being warm and welcoming to someone who was from "elsewhere." I lived alone in a tiny cabin near the ocean, and each night I'd walk up into the dunes and beam love out into the universe. In the re-telling, this may sound downright foolish. I hoped I wasn't the only one doing this; I trusted that there were (and are) many kind-hearted people standing under the stars offering love to other life.

Up until I wrote *Song of Increase,* I was a party of one. I sought to understand how our species could work together. Beyond a handful of close friends, I had no idea if anyone was interested in a book about my relationship with the bees. But I wrote it anyway.

What I found was a resounding *yes.* There were people interested in connecting with the bees. We are many, we just didn't know how to find each other. We have been brought together—you, me, and the bees.

That is why this very book has come about. We want loving kindness to become the standard for our relationship. We want respect for the bees' own wisdom to become a viable choice for bee tenders. We want to support more of us stepping away from conventional practices that treat colonies as chattel and rising up in our bee advocacy.

Don't waver. There are many, many of us. We have come remarkably far in just a few years, and with every person who makes the choice to help bees live the way nature intended, we make a reverent difference. In taking this path, our actions bless all life.

Acknowledgments

Susan: Jacqueline and I have a rare ability to write together in true literary partnership. One of us drafts a chapter and passes it to the other, and we go back and forth like this, our voices blending into one. I cherish this connection and this friendship so deeply. Together as friends, working as one, we have written for you this homage for our beloved bees.

I am grateful to my generous husband, John, who lets me vanish for weeks on end, both in body and in mind, when I'm writing, and for being gracious when a call about a bee swarm sends me bolting away from the dinner table.

I thank a host of close friends who helped me with my bee dreams and goals. Pixie La Plante has gone on many swarm calls with me, always lending a fine hand. Deborah Nagano has one of the clearest minds I know and has been an invaluable help with my bee dreams. She created the amazing array of functional bee art that graces our website bee shop.

My next-door neighbor Eric Larsen is a gifted artist with bee skills. Eric has helped me gather swarms and he created the illustration for our book cover. We asked him for a bee image that would entice people to fall in love with them. He pored over thousands of images and produced a drawing of sacred beauty. Now, I see the face of this beautiful bee in my dreams. The interior illustrations come from Jacqueline's talented hands.

To my European friends, Ferry Schutzelaars and Mike Albers, and all in the weaving community who have patiently worked with me and beside me, and to all those whose bee work is changing the world, thank you from the depths of my heart.

To Torben Schiffer and Michael Joshin Thiele, whose research and work has helped me to truly know bees, I am deeply grateful. You have been more than just friends and mentors. You are kin.

Jacqueline: My husband, Joseph, often clears his decks and lets my projects flutter up above his. He has a quick mind and frequently comes up with creative solutions to my wild ideas. He also has a strong back and a good sense of humor. How suited he is to matrimony.

A wave of gratitude goes to Robin Wise, who is an astounding wellspring of knowledge, from technical to spiritual.

To friend and author Celine Locqueville for her insightful book, *Ruches Refuges*.

I am in awe of John Phipps, who has championed the natural world of bees with every issue of *Natural Bee Husbandry*.

Special thanks to Janine Kearns for holding herself as a template of peace in the world.

Resources

Following is a list of books, DVDs, websites, and articles to deepen your knowledge and appreciation of the honeybee.

BEE-APPROPRIATE HIVES

We encourage you to explore these websites. They are full of innovations and possibilities for bee-friendly hives.

Kelsey Love (Sun Hive instructor): http://www.heirloomista.com/

Barry Malmanger—wooden inserts for box hive, plus his bee-friendly hive, the "Habitat Box.": barry11@jps.net

Torben Schiffer: https://beenature-project.com/Ueber-uns

Matt Somerville: https://beekindhives.uk/

Michael Joshin Theile: https://apisarborea.com

BOOKS

Bresette, Jack. *Sensitive Beekeeping.* Great Barrington, MA: Lindisfarne Books, 2016.

Bush, Michael. *The Practical Beekeeper: Beekeeping Naturally.* X-Star Publishing, 2011.

Fearnley, James. *Bee Propolis: Natural Healing from the Hive.* London, UK: Souvenir Press, 2001.

Freeman, Jacqueline. *Song of Increase: Listening to the Wisdom of Honeybees for Kinder Beekeeping and a Better World.* Boulder, CO: Sounds True, 2016

Frey, Kate, and Gretchen LeBuhn. *The Bee-Friendly Garden.* Berkeley, CA: Ten Speed Press, 2016.

Hauk, Gunther. *Toward Saving the Honeybee.* Junction City, OR: Biodynamic Farming and Gardening Association, 2008.

Heaf, David. *The Bee-Friendly Beekeeper: A Sustainable Approach.* West Yorkshire, UK: Northern Bee Books, 2010.

———. *Treatment-Free Beekeeping.* IBRA & NBB, 2021.

Kornberger, Horst. *The Global Hive.* Hamilton Hills, Australia: School of Integral Arts Press, 2012.

Lazutin, Fedor, and Leonid Sharashkin. *Keeping Bees with a Smile.* Ithaca, NY: Deep Snow Press, 2013.

Locqueville, Celine. *Ruches Refuges.* Stuttgart, Germany: Ulmer, 2020.

Seeley, Tom. *Honeybee Democracy,* Princeton, NJ: Princeton University Press, 2010.

Tautz, Jurgen. *The Buzz About Bees: Biology of a Superorganism.* Berlin and Heidelberg, Germany: Springer Verlag, 2008.

The Xerces Society, *Attracting Native Pollinators.* North Adams, MA: Storey Publishing, 2011.

DVDS, MP3S, AND VIDEOS

Alternative Beekeeping Using the Top Bar Hive and the Bee Guardian Methods, Corwin Bell, available at BackyardHive (BackYard Hive.com).

Queen of the Sun, award-winning and visually stunning documentary directed by Taggart Siegel about bees and beekeepers who are working in bee-centric ways.

Videos by Jacqueline Freeman (SpiritBee.com). Instructional, relational, and entertaining videos of her life with bees, animals, and nature.

Videos by Michael Joshin Thiele. Thiele always engages bees with respect and deep appreciation. Visit his website, Apis Arboreal (apisarboreal.com), to see his latest videos.

Videos by Robin Wise, a bee-loving professional audio engineer who makes exquisite recordings inside the hive of colonies doing tasks.
Comb Building: www.spiritbee.com/mp3s/comb-building-audio-mp3
Bee Meditation: www.spiritbee.com/mp3s/bee-meditation-audio-mp3
Song of Increase: www.spiritbee.com/mp3s/song-of-increase-mp3

WEBSITES

We encourage our readers to visit our website WhatBeesWant.com for videos, blog articles, resources, and information on upcoming classes and events.

Jacqueline's website, Spirit Bee (SpiritBee.com), offers movies, audio files, and images of colonies singing the Song of Increase and reflecting other fascinating states of being in the hive, including a good audio file to listen to during a bee-medicine session. Sound by Robin Wise.

Susan's website, American Skep (americanskep.wordpress.com) chronicles her adventures making and keeping in straw hives.

BackYardHive (BackYardHive.com) contains information and hive technologies that encourage and enable backyard beekeepers to be successful, founded by Corwin Bell.

The Barefoot Beekeeper (biobees.com) offers information about practical, balanced, treatment-free beekeeping in top-bar hives.

Apis Arboreal (Apisarboreal.com), founded by Michael Joshin Thiele delves deeply and expertly into the topic of rewilding bees, and of tracking wild bees living in local watershed areas.

Holy Bee Press (HolyBeePress.com), a crossroads of honeybee conversation and bee salon, founded by Deborah Roberts.

Natural Beekeeping Trust (NaturalBeekeepingTrust.org) is a nonprofit that focuses exquisite attention on the needs of bees. On this website, you can subscribe to the *best* bee husbandry magazine on the planet: *Natural Bee Husbandry*. Their recipe for homemade food for your bees, referenced on page 147, can also be found at naturalbee keepingtrust.org/post/2017/03/03/march-newsletter

College of the Melissae: Center for Sacred Beekeeping (College oftheMelissae.com), Laura Bee Ferguson, director. Laura's college has wonderful information on the history of women beekeepers and oracles of ancient times.

Michael Bush, the Practical Beekeeper (BushFarms.com/bees), has extensive articles on every aspect of healthy, treatment-free beekeeping.

APPENDIX 1. SPECIES PROTECTION

The True Price of Honey: Species Protection for Honey Bees!

Torben Schiffer

This article deals with the protection of species and the preservation of the ecosystem-relevant key species of honey bees and wild bees, which make an invaluable contribution to the preservation of our ecosystem due to their pollination activity. This is how the more highly developed flowering plants arose around 120 million years ago together with their pollinators, the solitary bees.

With the appearance of colony-building honey bees around 45 million years ago, numerous other flowering plants evolved, which to this day—in particular through the formation of fruits—constitute the basis of life for countless species. This "lock-and-key" relationship between flowering plants and bees has been instrumental in maintaining the ecosystem that surrounds us.

Species protection is therefore far more important than beekeeping, which focuses on the economic exploitation of honey bees with all its manipulative methods and procedures. In the meantime, beekeeping not only threatens the continued existence of honey bees themselves, but also deprives the cultural landscape in Germany alone of some 100,000 tons of nectar each year, only to harvest just under 30,000 tons of honey in the end.

Remarkably, many beekeepers complain about the shortage of nectar and regret what they regard as the only alternative: cruel treatment of their bees with acids, brood culling, and other means in order to curb the Varroa mite population. The fact that it is these established beekeeping methods that ultimately produce these side effects—and even threaten the continued existence of the species in the long term—does not seem to be an issue or to be clear to many.

Livestock farming officials often reject an open and objective discussion at the specialist level. Nevertheless, in the years to come we will not tire of waging these battles, because we owe it not only to the bees, but also to future generations. If we want to preserve the earth and the ecosystem that surrounds us for future generations, then we cannot avoid a paradigm shift in many areas, including beekeeping! It is time to put the established beekeeping lobby to the test and separate its romantic image from reality on a factual basis.

The public has a right to know how bees are treated in modern beekeeping, because we all depend on the survival of this systemically important key species. We are all part of this ecosystem. But the beekeepers themselves have a right to be shown the urgently needed and overdue alternatives to the established ways of operating, because many are not interested in honey at all: for them, it's only about the bees.

I do not intend to overtly attack or criticize the beekeepers personally. I am talking about data, operating methods, breeding criteria, and their significance or effects on the bees and the environment. I talk about systemic errors, conceptual problems, and lobbying. Pointing the finger or defaming individuals is not part of my character! I have given numerous lectures all over Europe—all to a professional audience. The people I met have in common the fact that the vast majority love their bees and only want the best for them. At the same time, this very majority suffers by their own hand, because the bees are like long-term drug rehab patients: as soon as medication is stopped, the animals die in apocalyptic proportions. The treatment of the Varroa mite and the constant interventions are undoubtedly a challenge for every empathetic person. Many aspiring beekeepers therefore drop out of their training at an early stage, and those who finish it often have a persistent, diffuse malaise.

All those idealistically motivated animal and species conservationists who originally only wanted to do something good for bees and nature were ultimately deprived by the establishment of the freedom to choose whether they even practice intensive livestock husbandry—including the associated side effects—or simply just want to do something good for the bees and nature. "Modern beekeeping," with its frames, numerous manipulative interventions, expansion of hive volume, breeding, and standardization, is widely propagated and taught to be the only way to keep bees. It is this operation modality that creates the numerous problems that ultimately have to be treated with medication.

In the end, it's not about the bees themselves, but about business. The beekeeping associations want to acquire members and pay salaries; the equipment sellers want to sell numerous devices; the pure breeders want to sell their supposedly gentle, high-performance bees; the pharmacy their drug catalog; and even the government bee research institutes get their subsidies in order to support the many projects "researching" the problems around keeping honey bees in hives, and to present solutions. All of these institutions earn their money within the current system.

Ultimately, only a few individuals determine the comprehensive training content and ways of dealing with the bees in beekeeping. However, the specifications made were left behind by the current *zeitgeist* years ago, and can now unquestionably be regarded as lobbying.

To illustrate by analogy: If I wanted to keep chickens out of sheer love for the animals themselves, then I would not let myself be persuaded that I have to complete training as a certified intensive poultry farmer; or that I need all the equipment necessary for this type of husbandry; nor would I cram the animals into a cage and then use medication to combat the husbandry-related side effects . . .

Are there really alternatives? Of course! Modern beekeeping is a modern problem. In the historical context, the current form of husbandry is less than the blink of an eye in its millennia-old history.

The current problems of intensive animal husbandry are by no means surprising, but were described quite precisely over 70 years ago in a book by Johann Thür. Even then, there were many warnings against the current operating modes and their side effects. All of the almost prophetic predictions made at that time have now become reality.[1]

ONE OF THE MOST IMPORTANT SPECIES ON EARTH HAS BEEN MONOPOLIZED IN INTENSIVE, DRUG-SUPPORTED LIVESTOCK FARMING

How could things have gotten to this point? On the one hand, it has always been about increasing efficiency and yield. In this area, beekeeping has made tremendous progress over the past few decades. Beekeepers were still harvesting an average of 10 to 15 kilograms of honey per colony per year as recently as a few decades ago, today it is 40 to 60 kilograms, sometimes even significantly more. These superlative results are achieved through a multitude of manipulative interventions, and by keeping bees in large hives or expandable magazine hives, all of which have volumes (and operating modes) that can no longer be reconciled with the natural and species-appropriate lifestyle of bees living in tree cavities.

On the other hand, mankind has changed the earth significantly in the last 75 years. After the Second World War, there were just 2.31 billion people on earth in contrast to the nearly 8 billion people today. In addition, large parts of Europe were in ruins. Wood was also needed as a building material for the reconstruction. Even the reparations payments to the victorious powers had to be made partly in wood. The bottom line is that 10 to 15 times more trees were felled than could grow back. Many of the ancient habitat trees in Europe fell victim to this movement. Nowadays, almost all animals that had adapted to life in tree cavities in the course of evolution have become rare, or are already on an endangered list.

So, there were two correlating movements. One can be described as "Industrialization 2.0," accompanied by explosive population growth. Higher, faster, further, more efficient... This philosophy was also transferred to honey bees, among other things. The other is characterized by the over-exploitation of nature.

The large-scale elimination of natural tree hollows led to another significant loss: that of a natural balance in the form of a gene pool existing predominantly in wild honey bee populations, so that modern beekeeping now has a *de facto* monopoly position on one of the most important species on earth. While around 570 wild bee species have already been placed under strict species protection, the honey bee alone is unlawfully denied the protection of species to which it is entitled. In the Federal Species Protection Ordinance, it can be read that honey bees only occur in domesticated form, and are therefore excluded from the protection ordinance.

When this claim was written into law, there was not a single scientific study that could confirm this statement. However: recent studies show that honey bees are not extinct in our forests, and that they do not share the systemic diseases of their managed cousins (foulbrood, varroosis, nosema, etc.), which are, of course, the consequences of this factory-farming style of beekeeping! The latter is also reflected in numerous scientific studies carried out on the topic.

THE COLLATERAL DAMAGE OF BEEKEEPING OR "BYCATCH", AN ENVIRONMENTAL DISASTER THAT HAS BEEN IGNORED SO FAR

The nature conservation organizations have wrongly lost focus on the protection of honey bees, because the current form of increased efficiency in honey production deprives numerous threatened species of their livelihood in the form of nectar. The exact dimensions of this depend on many factors (such as the distance to forage source, position, type of hive, volume, weather, etc.) and is therefore difficult to calculate. Exemplary calculations, which are based on known scientific data, reveal the historically unprecedented (and so far neglected) over-exploitation of the widespread intensive animal husbandry of honey bees, which is taught as the only way.

Professor Jürgen Tautz calculated—from the energy expenditure in the hive, the energy content of the honey and number of collective flights—that a colony of bees in a Langstroth/National hive produces up to 300 kg of honey per year, of which by far the largest proportion is burned as heating fuel.[2] If we assume for the calculation that 50 kg of honey can realistically be harvested by the beekeeper in the course of the year, then up to 250 kg alone would account for the basal metabolism.

This base level of energy, which the bees burn—unnoticed by the beekeeper—in the background for thermal regulation and comb building, can in turn be extrapolated to a corresponding amount of collected nectar. A colony of bees needs 3 to 4 liters of nectar to produce one kilo of honey (depending on the sugar content). According to this, a conventionally kept bee colony in a large hive would consume between 750 to 1000 liters of nectar per summer just to maintain the vital core heat and to build its comb in this inappropriate, oversized hive. For the currently 900,000 bee colonies that are kept in this way in Germany, that would be between 675,000 and 900,000 tons of nectar that we extract from nature—every year! Even if the basal metabolic rate were calculated with only half of 100 to 125 kilos of honey per year, the nectar quantities required for this would still reach amounts of up to 400,000 or 450,000 tons per year—in Germany alone.

AN INCONVENIENT TRUTH

How many thousands of tons of wild bee populations (biomass) could be produced from the amounts of nectar used in modern beekeeping? How many second- or third-order species could live on this excess? The "bycatch" of modern beekeeping undoubtedly removes a large proportion of the nectar from the ecosystem of the cultivated landscape, which was available for millions of years to the entire diversity of pollinator insects and the food chain based on it.

The basic metabolism in standard hives exceeds the energy requirement in tree hollows by a factor of ten. The basic principle is simple: Everything that goes out has to be brought back in! The largest part of the total amount that a bee colony brings in per year is used for the internal processes. Thus, a large part of the brood and the associated Varroa mite population in beekeeping is only produced to combat the loss of heat energy. The quantities of nectar removed from the cultivated landscape have a direct effect on the quantity of hundreds of endangered pollinator insects. The only comparable such waste is observed in the fishing industry (bycatch).

Ultimately, the many blooming plants do not produce their nectar so that we humans—with the help of this widespread, heavily-intensive honey bee factory farming—can divert it into a glass jar. Ironically, the majority of this is used up exclusively in compensating for the inappropriate conditions found in modern beekeeping. The plant energy extracted from the system of the environment in the form of nectar has a direct effect on the quantity of numerous insects, birds, bats, hornets, and other species. Finally, the nectar serves as a food basis for various species. Food sources are known to be converted into living biomass. The amount of nectar or honey taken from the cultivated landscape influences this sensitive balance. This natural law is based on basic biology and is easily understood. Today, however, these principles apply more than ever, as the problems have increased due to intensive agriculture and pesticides, as well as the loss of flowering areas, habitats, and climate change.

Interestingly, the working group of the Institute for Bee Research came to the conclusion that there is no verifiable competition between wild bees and honey bees for nectar:

> *Recent studies have also come to the conclusion that the presence of honey bee colonies does not endanger the existence of wild bees. From this it can be concluded that honey bees—at least in their traditional range—do not pose a threat to wild bees. In the natural areas of distribution, it can be assumed that honey bees and wild bees have evolved into a coexistence.*[3]

The last sentence in particular shows that modern aspects such as land use and beekeeping are not considered at all in this statement.

The interpretations listed here reveal that science not only serves to create knowledge, but can also lead to the realization that common sense is relative.

How many hectares of wildflower meadow would we have to plant to make up for hundreds of thousands of tons of nectar? When the landscape is "sucked dry", the bee colonies are fed with sugar water—but what happens to the wild bees? In some regions in Germany there was a decrease in the biomass of flying insects by 75% in the last three decades alone.[4] Of the approximately 570 native bee species nationwide, around half are already on the Endangered List, and around 40 species are already considered extinct.[5] This is an almost absurd situation when you consider that the enormous pollination performance of honey bees has positively influenced the quantity of flowering plants for over 45 million years, and thus also favored the diversity and quantity of various pollinator and other higher-order species.

THE BASAL METABOLIC RATE AND THE NECTAR SHORTAGE: IN MANY CASES A MAN-MADE PROBLEM OF MODERN BEEKEEPING METHODS AND INAPPROPRIATE HIVES

Bee colonies are forest creatures. This has several advantages. Through the several hundred liters evaporated daily by each tree (depending on its size), a stable microclimate in the forest is ensured, with the temperature peaks that occur throughout the day thus balanced out. In addition, the canopy protects the tree cavity from direct sunlight, wind, and precipitation. Further, the tree cavity is usually itself—due to the small inside diameter, the massive sidewalls, the special physical properties of the wood, and the wood morphology—extremely energy-efficient and therefore climate-stable. However, thin-walled wooden boxes are not sufficient for these climatic effects, because solid, thick-walled and open-fibre wood is required for this.[6] Two other factors are also of major importance: On the one hand, Thomas Seeley was able to prove in his experiments that bee colonies always choose a dwelling at height; on the other hand, his experiments showed that bee colonies prefer tree hollows with a volume between 30 and 60 liters over other volumes. These preferences are ultimately the result of natural evolution that has lasted millions of years. Bee colonies that colonized cavities close to the ground fell victim to natural selection just as much as those which chose too-large volumes.

Proximity to the ground and large volumes are therefore selective factors, which bee colonies mostly did not survive even before primeval times. In beekeeping, however, volumes are regularly used which can reach up to 200 liters. In addition, the boxes are placed low to the ground, or even directly on it, leaving these large-volume boxes directly exposed to weather conditions.

Siting beehives close to the ground also ensures more humid internal conditions due to the moisture in the soil. In addition, soil mainly consists of destructive agents such as fungi and bacteria that decompose organic material and form air-permeable spores.[7] This is one of the reasons why there is no colony-building insect in our latitudes that can survive the winter underground in a waking state, living off of stores. The fact that bees were able to stay awake, consuming their winter stores for millions of years is due to the fact that they lived in a tree hollow—and were protected from harmful microorganisms and the rigors of the weather by their antibiotic, propolis-filled air.[8]

> *"The combs are the dwelling assigned by creation to the bees, whether in the hollow tree trunk, or securely sealed to the walls of the straw skep; each comb alley forms a closed space, a room as it were. In winter, therefore, the warmth of the cluster cannot escape through numerous gaps between frames and hive walls: loss of heat, drafts, moisture and excessive consumption of stores are avoided.[. . .] It is just as clear that even in the most sophisticated hive, no matter how thick-walled it may be, the bees cannot thrive properly if the law of narrow cavity nest-scent and heat retention is not fulfilled."*
>
> —*Beekeeping*, Johann Thür, 1936[9]

Wall thickness of a natural tree cavity (left) compared with a standard hive. Not only do the thick walls influence the indoor climate, but also the diameter. In a tree hollow, the heat is concentrated in a small space and is well insulated by solid walls, while in the hives it is largely lost via the large, thin surfaces.

Hive location has a not-to-be-underestimated influence on the required basal metabolism and thus on the overall behavior of the bee colony. The temperatures in a city can be up to 10° warmer than in the country on a summer day. The forest climate is, on average, another 4° cooler. A bee colony in urban housing (set up sheltered from the weather) must therefore compensate for a temperature difference of 14° on a normal summer day, compared to natural living conditions.[10] In fact, the temperature differences are actually much higher, as most of the bee colonies live directly under open sky or are placed directly on flat roofs, and are therefore exposed to the rigors of the weather, especially wind and direct sunlight, as well as the summer heat. Due to the large volume of these hives, the thin walls, the physically unsuitable designs (e.g., large, slit-shaped entrance holes, screen floors, heat-leaking frames, flat, sprawling surfaces, large diameters, corners, etc.), the bees are constantly forced to compensate for temperature differences. The energy required for this is ultimately provided by the nectar.

The widespread intensive animal husbandry of honey bees is a historically recent phenomenon. Until a few decades ago, bee colonies were mainly kept in warm, single-chamber hives of limited volume, such as straw skeps, which were also sheltered from the weather, and set up in special stands or bee houses. Compared with today, these colonies only required a fraction of the basal metabolic rate.

The latest research suggests that a bee colony living in a tree cavity (compared to a colony in a standard hive) needs less than a tenth of the nectar energy for the basal metabolic rate per year.[11] Conversely, this means that a modern beekeeping colony exceeds the nectar consumption of ten natural tree cavity colonies. If the bee colonies were kept in a species-appropriate manner, hundreds of thousands of tons of nectar in Germany alone could be available again as a food source for other pollinator insects—now often edged out by honey bees.

Adding to this problem, there is a high density of colonies in many places. In particular, the hype about city beekeeping has led to, for example, situations in cities like Berlin, where there are around seven colonies per square kilometer.[12] In the worst case, up to 1000 liters of nectar (per colony) alone are necessary for basic colony metabolism if kept in inappropriate, large managed hives. Since these quantities are of course not available in the city, sugar water often has to be used as an emergency feed in summer after the spring nectar flow so that the colonies do not starve. In addition, 300 species of wild bees live in the capital, three-quarters of which are threatened with extinction. Intense competition for food—which occurs in many places due to modern honey production—ensures, among other things, that thousands of starving or already dead bumblebees can be found under the linden trees in the city as early as June. Nevertheless, the sale of the species is being pushed forward under the guise of a "good deed" for bees and nature. Start-up companies such as "There is a bee on the roof /Bee-Rent," or associations such as "*Stadt-bienen* (Citybees)" meanwhile partner with corporations, who provide financial support for this cruel method of keeping bees in this alien urban environment.[13]

BEEKEEPING IS ANIMAL HUSBANDRY, AND NOT A SPECIES CONSERVATION PROGRAM!

In beekeeping, it is primarily a matter of exploiting the bees' labor as efficiently as possible to obtain the product. In addition, all international criteria for animal welfare (the five freedoms of animals) as well as the current animal protection law with regard to the current mode of beekeeping are broken. With their promotion, numerous species threatened with extinction are further squeezed.

The problem is also intensified by targeted propaganda from associations such as "Mellifera." Here, terms such as "essential beekeeping" are used without even remotely doing justice to the essence of bees. In this way, the association acquires numerous idealistically motivated members who actually want to do something good for the bees and nature, and transfers them to the monocultural training of intensive animal husbandry. At the end of the day, the bees in "Mellifera" are kept just as inappropriately as in other beekeeping associations, so that even here, the bees do not escape painful treatments with acids and other treatments. So, the side effects of this "essential" method of beekeeping—which is propagated as "natural"—are indistinguishable from those in conventional beekeeping! At the same time, there is apparently a lack of understanding of little things, such as a species-appropriate housing, or the fact that all heat energy that is lost outside must be compensated for by the bees themselves.

A bee can only bring in a certain amount of nectar in the course of its life until its body or its wings are worn out.

Professor Tautz's calculations show that a bee can collect about 2g of honey during its foraging stage. The stores can thus be converted not only into bee lives, but also the energy losses of the hives themselves. The bees have to adapt their brood cycle to the hive's energy loss, along with the associated wear and tear on the foragers.

An increased loss of energy or overcompensation is paid for correspondingly with the life of the forager bees, which are all short-lived. The compensatory performance of colonies in large hives not only burns a large part of the input (for maintaining the basal metabolic rate in temperature control, brood-rearing and comb building), but also—when correspondingly adjusted—shows an increased rate of brood turnover (and Varroa mite reproduction) and in turn significantly determines the behavior of the colonies.

Energy-leaking, bathtub-style boxes are neither natural nor appropriate for the species compared to tree hollows. Nevertheless, these (in inferior-quality plywood) are sold in hardware stores, and are described by the Melliferas Board as literally "species-appropriate beekeeping for the self-sufficient". This is not only factually wrong, but could also be viewed as consumer fraud.

The efforts of the association's leadership to present the appearance of species protection to the outside world is a symbol of a self-serving organizational policy that is detached from the natural needs of honey bees. Unfortunately, the term "essential" has been used thus far as nothing more than an empty phrase. It is still very popular, because it is mainly associated with a seal of quality and with a special bee-friendliness, or even species protection.

The bees pay the price for these questionable depictions, but they neither show facial expressions nor do they have vocal cords—the suffering takes place in absolute silence!

Nevertheless, a clear distinction must be made here between the Board of Directors and the membership of the association. A bee colony does not only consist of the queen and her entourage. The numerous beekeepers of the association are characterized above all by the constant, open search for improvements in their way of beekeeping, which is mainly driven by the suffering of the bees in the current form of beekeeping. There can therefore be no "business as usual" in the future. The efforts of the grassroots are therefore often in direct opposition to the actions of the Board of Directors, who constantly glorify the current situation with flowery words. In addition, the established operating methods and hives are not sufficiently questioned. Instead, one warns against being treatment-free, and enforces the current state, which ultimately hinders the evolution of honey bees.[14]

IF WE LOVE BEES SO MUCH—IF WE REALLY CARE ABOUT BEES—WHY DO WE KEEP THEM UNDER SUCH INAPPROPRIATE AND CRUEL CONDITIONS?

In the German language, the word for "hive" is *Beute* ("loot," "booty," or "prey"). And this term describes the use of this geometric shape very aptly: it is primarily about exploiting the bees—turning them into something to be looted. Nothing in a hive was designed for the bees, but rather all aspects serve only the simplest possible, barrier-free manipulation and suppression of natural behavior. The hives have all been specifically designed for this purpose and designed to function like a production machine. A hive is therefore directly comparable to the confinement of every other livestock operation—e.g., pigs, cattle. In addition, the associated impressive range of standardizations and interventions (e.g., honey supering, frames, divider boards, swarm prevention, reproductive splits, forced combinations, etc.) has led to unprecedented efficiency gains in the yields that can be obtained. At this point I would like to recommend the text "Bees in trouble—or are we using the wrong beekeeping methods?" by Georg Peukert, which can easily be found on the Internet.

DIFFERENCES IN RESPONSIBILITY BETWEEN LIVESTOCK HUSBANDRY AND SPECIES CONSERVATION

Attempt to keep bees in standard hives and simply not treat them can be equated with the idea of stopping medication in factory farm stables, and watching the suffering and death that follow. This is therefore not just a violation of animal rights, but cruelty to animals. The fact that the animals fall ill and die in large numbers in such experiments is understandably not due to the animals themselves, but rather the conditions in which they have to live. So, following this simple principle, the side effects that occur systematically in modern beekeeping cannot be assumed to affect the colonies living in the wild! The proposition that "there can no longer be any viable wild colonies because the Varroa mites would destroy them" shows that there is a huge gap in knowledge about the biology and behavior of bee colonies in tree cavities. It shows further that, insanely, the conditions of bees under a livestock farming model have been equated with a life under natural conditions. One could just as well propose that wild boars must no longer exist, since their domesticated relatives are dependent on constant medication in intensive livestock farming.

If we keep (exploiting) animals of any kind in artificial systems, then we undoubtedly have a responsibility, especially in compensating for the side effects of the method of farming itself.

In this purposeful form of animal husbandry, every bee colony is kept alive through medication or supplementary feedings. Whether the bees carry the genetic information in them to be able to survive in nature independent of humans is completely irrelevant. The age-old processes of natural selection are eliminated. A dead bee colony is understood as a "beekeeping error." The success of this "livestock" approach is not only measured by the honey yield, but also by the survival rate of the colonies. The hive conditions, operating methods, and medications thus actively block the adaptation of the honey bees to their respective environmental conditions. In this way, their evolution has been frozen—just as they themselves are frozen in a state of permanent medication.

"A BEE WOULD NEVER GET INTO THE HONEY BUSINESS—IT WOULD COST WAY TOO MUCH." (Jonathan Powell of the Natural Beekeeping Trust)

In contrast to beekeeping, species conservation is not about "keeping livestock" or compensating for genetic weaknesses in bees through medication or other substitutions (or artificially keeping colonies alive for human purposes, thereby preventing evolution). It's not about beekeeping or looting! It's about providing the numerous swarms and other tree-cavity dwellers with a species-appropriate habitat that they have largely lost through the clearing of the forests, and thus giving them the best possible basic conditions for survival. The point is to free the bees from all criteria of livestock husbandry and to enable them to have a future independent of humans. It is important to let loose the evolutionary processes that are millions of years old in order to restore the natural balance to industrial animal husbandry and the associated genetic erosion. The death of the non-adapted individual (or colony, in this case) is a completely natural and extremely important evolutionary factor. Natural selection means that only the most well-adapted offspring can survive.

NATURAL SELECTION IN THE HIVE?

There is no "natural" selection under unnatural conditions (stopping the medication in intensive animal husbandry does not result in a "natural" selection and can at best be assessed as a breeding attempt). If we want to ensure that bees are able to survive independently of humans, then we have to offer them the living conditions that they've always found in tree cavities through the ages. After all, bee colonies that undergo selection in hives do not necessarily need to be able to survive in a tree cavity, and vice-versa. In our rewilding projects, for example, it has often happened that swarming bees—who had nicely drawn out their foundation in their previous box hive—were not able to build proper comb in their new tree cavity, and thus fell victim to natural selection within a few weeks. In this way, the existing gene erosion is cleaned up by nature itself and the non-adapted genetic material disappears from the gene pool.

The responsibility of the species conservationist is therefore completely different from that of the livestock beekeeper. A conservationist does not subdue the animals to be protected, but rather protects them. Bees don't belong in boxes! The brood quantities generated in intensive animal husbandry do not exist in tree hollows, nor the need for the bees to bring in unbelievable amounts of nectar to fill empty boxes or to compensate for enormous energy losses, nor the overpopulation of Varroa mites caused by swarm prevention. The behavior of the bees changes significantly as a result. Millions of hours can now be devoted to natural behaviors that make survival independent of humans possible.

The responsibility of the conservationist is to replace the lost habitat of the tree cavities and to protect the subsequent natural behaviors and processes. In this way, natural selection is allowed again, just as glorious as a naturally occuring swarm. And that is how we get to a completely natural population dynamic. In spring 2020 I started in the *Alten Land* with 13 bee colonies. Nine of them were bought from beekeepers and each put in a single deep box. Four already lived in species-appropriate conditions. Since there was no expansion or intervention, the bee population grew via swarms to a total of 36 colonies by August. If 22 of the colonies die again by the coming spring, the population would again reach the initial size of 13 colonies and would therefore be stable. Because only the adapted bee colonies survive, swarms that are better adapted will colonize the empty tree hollows in the coming year. This pulsation of population dynamics is the natural and absolutely normal state. Only through this principle has there been a constant adaptation of the species to the constantly changing living conditions of the environment since primeval times, and only these processes ultimately guarantee the long-term survival of each species.

BREEDING, GLUTTONY, AND MEDICATION—THE INSTITUTIONAL ABOLITION OF EVOLUTION

Contrary to the statements of some "conventional" bee researchers, who claim that the honey bees are not endangered because the beekeeper takes care of them, the honey bee species is being driven further to extinction, as humans intervene at will in the 45 million-year-old gene pool of honey bees in order to be able to "manage" them in an even more relaxed and thorough manner.

What is overlooked is that any bee behavior selected for also has a cost that is paid somewhere else. The criteria that are regularly expected from honey bees in beekeeping, such as docility, honey production, stable comb, reluctance to swarm, and low propolisation all have in common the fact that each reduces the survival probability of the species under natural conditions. The targeted breeding and selection of honey bees for properties desired by humans therefore not only threatens the species itself, but also represents a long-term danger that should not be underestimated for the entire ecosystem in which we live and of which we are part. For millennia, the bees' biology and thus their evolution were largely left untouched. Even breeding was initially irrelevant for the survival of the species as a whole, as long as the predominant part of the bee colonies was located in nature. The honey bee colonies living in the forests are subject exclusively to natural selection and adaptation.

Beekeeping can now look back on a culture that is thousands of years old. Ever since humans have kept bees, they have always made use of natural bee colonies that have been adapted by natural selection. However, this source of vital, adapted, and viable genetic material has almost dried up.

The relationship between the bee colonies subject to livestock husbandry and those living in the wild has changed dramatically. Although 31% of Germany is forested, only 6% is conservation land. In total, Germany has a forest area of around 107,000 km^2. The latest research shows that bee colonies in the forests are by no means extinct. However, colony density is low. If we assume a positively estimated density of maximum one colony per square kilometer of forest, then there are around 100,000 wild bee colonies in Germany compared to 900,000 in livestock husbandry. The gene pool of these ancient key species is therefore no longer subject to natural selection, but to human whim.

The breeding criteria of the German Beekeeping Association show particularly impressive concepts of "race-breeding" or "pure-breeding." Bees are not only bred for the well-known beekeeping properties such as honey production, docility, colony strength, lack of swarming instinct, etc., but also for aesthetic characteristics such as coloring, hair length, striping, and cubital index (a fine line in the branching of the wings). Bees that do not correspond to the desired aspects of their self-empowered "creator" are designated "unsuitable bee material".[15] Human capriciousness decides here whether a bee colony is "worth breeding" and not the property of being able to survive without human intervention in nature. The breed criteria are completely artificial, as nature does not ask the bees for breed names, hair colors, or aesthetic criteria—only properties!

Here is a recent example: Thomas Seeley was able to prove in his scientific analyses that the wild honey bees changed genetically after the arrival of the Varroa mites. The species was thus adapted to the existing environmental conditions through natural selection. The honey bees that live in the woods around Cornell today are not just genotypically and behaviorally varied, but have also changed phenotypically (in terms of appearance): The workers have a smaller head circumference and thorax (middle body) diameter as well as a different wing shape than those who lived there before the arrival of the Varroa mite. One should rightly assume that no one would come up with the short-sighted idea of keeping the "old" honey bees, who have fallen victim to natural selection, alive through protected breeding methods in order to save them from "extinction". Idiotically, this is exactly what is practiced in the pure breeding clubs. These biologically and ethically incomprehensible breeding criteria would—due solely to their minutely defined criteria for visual aesthetics—not give rise to the kind of adaptations that were seen in the Arnot Forest after the arrival of Varroa.

The gene pool must be in constant flux in order to remain viable, and must not be changed on a large scale by mindless breeding methods, or blocked by industrial, intensive animal husbandry.

The traits changed by humans will perhaps still be detectable on a genetic level in thousands of years, but strictly speaking it is only a matter of the compression of already-existing genes (recombination). Therefore, in no way can any kind of "claim to possession" of the species be asserted. The widespread one-sided view of so-called pure breeding does not do justice to reality and the actual prevailing conditions in any way, and must be reconsidered—in particular due to the fact that the predominant gene pool of contemporary bees is in human hands and its natural selection is therefore undermined.

In this context I would also like to emphasize that honey bees are wild animals! The discussions about whether they are domesticated pets or wild animals have flared up again due to the current investigations into wild, viable colonies. Far too often I've heard the argument that honey bees are now only highly-bred, high-performance bees. These relationships must be considered in a factual manner! Bees have existed for about 45 million years, breeding for only about 100 years. In breeding, we condense certain already-existing genes, but do not create them anew. So it happens that the recombination of genes that takes place during reproduction causes the original behavior (for example defensiveness) to break through again and again. The lower the proportion of genes in the gene pool that are formed by natural selection for adaptability and survivability, the less often original behaviors occur. For this reason in particular, the species is likely to continue to lose its ability to survive independently of humans if things continue as before.

If bees were indeed already domesticated pets, why do we have to take apart the bee colonies at regular, close intervals to disrupt their natural behaviors? Why do the bees still want to swarm? Why do the scouts still reliably find every potential dwelling in the area? Why do we need protective suits, smoke, water sprayers, etc. to "work" with the bees? When I pet my dog on the sofa in the evening, I don't need a full-body protective suit, fog, or water cannon, because the dog is undoubtedly a domesticated pet. If I tried the same thing with a wolf, however, I probably wouldn't be able to do so without proper equipment.

IF IT WEREN'T FOR HONEY, THE HONEY BEE WOULD BE PROTECTED

The biggest problem for honey bees is not the Varroa mite, but the extensive, evolution-destroying subjugation of this key ecosystem species through the intensive animal husbandry of modern beekeeping, which violates animal rights. Although the majority of current hobby beekeepers—and young beekeepers as well—keep bees for idealistic reasons (or have had a change of heart), they are currently still being shunted into this monocultural training system for intensive animal husbandry. Even the governmental bee research institutes play a major role in this.

In June 2020, the Institute for Apiculture in Celle published an article with the title "Animal Welfare (Bee Welfare) and 'Good Beekeeping Practice.'" In particular, this text serves the conventional view of bees as production machines, and focuses on combating the side effects of modern forms of husbandry that are depicted as having no alternative. Interestingly, the text correctly points out that paragraphs one and two of the Animal Welfare Act also apply to bee colonies.

§ 1 Animal Welfare Act: *The purpose of this law is to protect the life and well-being of animals, recognizing our human responsibility for them as fellow creatures. No one may cause pain, suffering or harm to an animal for no good reason.*

A CONVENTIONAL OPERATING MODE IN WHICH THE FOLLOWING ACTIVITIES, AMONG OTHERS, ARE CARRIED OUT:

- the wings of the queens are clipped so that they cannot swarm
- bees in completely alien (selective) conditions, in expandable boxes e.g., made of styrofoam, close to the ground and far too close together (crowding factor)
- brood chambers can be expanded to around 80 l (two deeps) as standard

Large amounts of brood lead to large amounts of Varroa mites. Swarm prevention increases this problem. In addition, the amount of brood is only generated so that there is enough "workforce" to fill the "standard" 80 l empty space with honey. Volume alone is a selective factor.

- constant addition of empty space

The colony always tries to fill space in order to achieve the vital security of stores. The millions of working hours necessary for this come directly at the expense of natural behavior.

- the bee colonies are divided into frames

Frames prevent natural behavior in a variety of ways. On the one hand, there is a disturbed heat balance, as the heated hive air constantly flows away into the large space due to bee-space gaps and is lost. On the other hand, the natural formation of the bees in three-dimensional nets, which are made up of chains of bees (visible in every tree cavity), is prevented. This formation of chains serves, among other things, climate stabilization, communication, and the defense against invading enemies.

- the bees are treated with smoke once a week in order to control the entire hive or to check for queen cells, during which crushing bees usually cannot be avoided . . .
- queen cells are deliberately crushed to death in order to "abort" natural reproduction . . .
- drone brood is cut out in order to contain the Varroa mite population, which is, of course, created by this mode of operation in the first place

This practice means that thousands of drones slowly die in the cut-out honeycombs.

- Almost all of the honey is stolen from the bees and replaced with nutrient-free sugar water . . .

Research clearly shows feeding bees sugar water leads to pathological changes in the gastrointestinal tract, disrupts metabolism, and ultimately leads to a shorter lifespan due to malnutrition.[16]

- the bees are assaulted with caustic acid fumes and wounded . . .

The standard formic acid treatments kill portions of the brood, in some cases even drives the bees to tear out their antennae, and in some cases kills entire colonies

- the colonies are housed in the physically unsuitable geometries (hives / boxes) that exhibit stark differences from natural living conditions in tree hollows . . .

In winter in particular, mold regularly forms on the storage combs, which endangers the health of bees. The bees have to survive for a full six months under these damp, moldy conditions with nutrient-deprived sugar water as food. Other factors such as proximity to the ground, unnaturally large volumes, and heat loss also apply.

- Breeding by humans

Artificial insemination, single-drone insemination, breeding away of vital natural properties (e.g., defense behavior) for easier exploitation.

How can it be that an operating mode that denies the evolutionary rights of the bees (all beekeeper interventions ultimately serve to break the natural behavior of bees) and also violates international guidelines (pain and suffering, malnutrition, mutilation, chemical burns, inappropriate housing and installation, targeted killing of drones and queens), be described as "animal welfare" and "good beekeeping practice"?

If by law no human being is allowed to inflict pain or suffering on an animal without good reason, then this modern form of keeping bees is illegal according to my understanding. Profit maximization is certainly not a reasonable reason to deny bees or other (wild) animals any integrity, any right to physical integrity and the exercise of natural behavior and, moreover, to keep them in the most questionable conditions (violation of §2)!

In this form of dedicated livestock husbandry, the bees are reduced to the status of a maintenance-intensive honey machine, on which one has to constantly intervene and manipulate so that it runs properly. Today's beekeeper education is no longer about the needs of the bees themselves! Rather, it is about the controllability of the many various interventions, yield maximization, and—ultimately—battling the side effects of this industrial factory farming, which has nothing to do with natural aspects.

If we also look at how many hundreds of items there are to buy in the well-known larger beekeeping shops and what these are ultimately used for, something quickly becomes clear. Not a single product available there serves the bees themselves: the bees do not need any of it! All articles serve solely the manipulation, breeding, and exploitation. From the bees' perspective, this industry is likely to be a cabinet of horrors.

The article from Celle has further interesting remarks: On the one hand, the text rightly describes a swarm of bees as a wonderful natural phenomenon, on the other hand, it is regretted that swarms would rarely find suitable nesting opportunities in today's man-made environment.

One would imagine the functionaries of the institute should be very grateful for my work, because we hang tree-like bee-appropriate habitats (Schiffer Trees) in many places in order to counter this lack of species-appropriate habitat. Every year there are thousands of swarms that fail to find adequate housing. With artificial habitats that mimic natural conditions, we can put an end to this suffering, and also increase the proportion of bees living in the wild. At the same time, these colonies can even be easily monitored, as their location (in contrast to the current state) is known. Therefore, replacing the lost tree hollows is a long-overdue step that we have long followed for numerous other threatened animals (bats, hornets, bumblebees, birds, even wild bee hotels, etc.)

However, Celle believes that the health of bees—in contrast to other "pets"—cannot be investigated so easily. Furthermore, the "Schiffer Tree" (tree-like bee habitat) is literally referred to as the "Schiffer Hive," which would contradict the Animal Health Act and the idea behind it, as the colony's health cannot be ascertained due to a lack of intervention and invasive observation. The article continues to claim that untreated, wild bee colonies not only break the law, but also encourage the spread of disease. Really? Is that really based on facts, or is it a more "political" statement?

What do the real studies, that have achieved globally comparable results according to scientific criteria, have to say?

AMERICAN FOULBROOD (AFB)

American Foulbrood is one of the most dangerous bee diseases of all. This is a highly contagious bacterial infection that requires notification of the authorities. The spores of the bacterium themselves contaminate the honey stored in the honeycomb and also collect in the wax. The nurse bees transfer these pathogens through the larval food to the brood, which is killed and decomposed by the pathogens. This creates up to 2.5 billion new spores per larva! The colonies, which collapse due to the lack of offspring, are often robbed out by other bee colonies, which means that the disease spreads quickly.

In numerous historical articles it is noted that the disease began to spread at a worrying rate at the same time as the change in beekeeping methodology from nature-oriented beekeeping in skeps or the like, to frames and boxes. The reasons for this are not only the alien living conditions with which bees living in boxes have to deal, but also the mobility of the frames and the associated fact that combs can be swapped back and forth between colonies. In addition, the trade in bee products such as honey or wax, as well as equipment, entire colonies, or queen bees has increased significantly, so that material contaminated with spores can be distributed over long distances within a very short time.

For this reason, officially restricted areas are established when the disease occurs, in which all bee colonies are officially checked. The veterinary office can order the destruction of the infected colonies or a rehabilitation of the apiary. The strictest requirements must be observed here. All materials must be disinfected according to regulations and the contaminated honeycomb incinerated.

The fear of this serious, in some cases existence-threatening disease is understandably very widespread. The fear of beekeepers regarding wild bee colonies is correspondingly high. Historically, however, it has never been any different. Of course, colonies in "Schiffer-Trees" are also eyed suspiciously for the same reason, although monitoring is in principle possible here too.

In beekeeping, so-called brood-ring samples (wax and honey) are taken from the hives and examined for foulbrood spores in the laboratory as the standard for prevention. If the latter can be determined, the beekeeper can take targeted measures to prevent the disease from breaking out. Despite all these measures, however, numerous outbreaks and restricted areas occur every year. This is due in particular to the fact that it is a symptomatic disease that only arises and spreads due to the unsuitable keeping conditions in beekeeping. This shocking fact is proven by numerous historical as well as current scientific studies.

A study of all previously known cases of foulbrood diseases in beekeeping and the connections to wild bee colonies led to the realization that it is the conditions under which bees are kept that are particularly responsible for the diseases. Based on unambiguous data, the study convincingly shows that a threat to production hives from wild colonies is more than unlikely.[17] For this purpose, historical and current studies were combined. Some are listed here as examples.

In the US state of Michigan, a persistent foulbrood epidemic led to the passage of a drastic 1929 law that declared the numerous wild bee colonies in this area to be illegal. A large number (approximately 300) of these colonies were then exterminated. Remarkably, AFB could not be detected in any of them, while 13.3% of the beekeeping colonies in the same area were infected.[18]

In England, G. Wakeford (Sussex) examined 100 wild bee colonies over several years for AFB in the middle of the last century. Again, not a single case could be found, although the disease was present in the surrounding apiaries.[19]

A more recent study from New Zealand was able to identify the foulbrood pathogen in 12.5% of all tested bee colonies in apiaries (with fewer than 50 colonies). In wild colonies from the same region, however, spores were only detected in 6.4% of the 109 colonies examined. Further, the spore load in those wild colonies was very low in comparison to those in the apiaries. This is important to note, as a minimum level of infection must be reached before clinical symptoms (i.e., disease) become evident.

This leads to the conclusion that wild colonies are more likely to be endangered by managed colonies than the other way around. Thousands of cases of foulbrood diseases in apiaries should be contrasted with only three cases of this disease ever found in wild colonies: two in England, one in Australia. A dead bee colony was discovered in 1957 in a roof in Dorset near Brigadier (England). Records from the National Agricultural Advisory Service recorded 38 reported cases of American foulbrood in the same region between 1950/51, 16 of which were near the dead colony. Another infected colony was discovered in a roof near Dorset, England, and again, data collection by the National Advisory Service shows that 18 cases were detected in the village's managed hives during this time.

In 2002, Thomas Seeley investigated wild bee colonies in the Arnot Forest near Cornell University, in the United States. Swarm traps were set up, and 100% of the colonies examined showed Varroa mites, but no American or European foulbrood could be detected.

These data results should, by themselves, be reason enough to rethink the established housing conditions and make them more suitable for bees. The evidence that bee colonies get sick because of the way they are kept is overwhelming and more than clear. So how is it that the bee research institute in Celle comes to such a striking statement that wild bee colonies promote the spread of diseases?

Furthermore, the article "Good Beekeeping Practice" claims that the feral bee colonies would die of the Varroa mite after "a few years" anyway. I recommend that the functionaries of the institute look into current research reports by Thomas Seeley, Benjamin Rutschmann, Barbara Locke, etc., or to simply expand their "livestock science" to include studies on wild colonies. There are already numerous scientifically proven cases in which honey bees survive without any treatment.[20] I myself monitor several wild colonies that have been living for years without any interference.

It becomes clear how little the article's authors have dealt with the topic, and how little they know about the biology and behavior of wild bee colonies (they do not even succeed in distinguishing between box hives and tree-like conditions, or between livestock husbandry and species conservation). Perhaps this is also the reason why the article emphasizes the need to thoroughly learn how to keep livestock and, at the same time, only requires the trainees to have "basic knowledge" of honey bee biology. The authors continue to write that "beginners" who start with "Schiffer-Trees" would give up prematurely. Have they tried to run a conventional beekeeping with a tree-like cavity?

Let's summarize this briefly: When the management of a governmental bee research institute literally calls intensive beekeeping "Animal Welfare (Bee Welfare) and 'Good Beekeeping Practice,'" yet describes natural selection as a violation of animal rights, then this raises many questions! What exactly is the purpose of these institutes? "Livestock science" has been practiced *de facto* for decades, whereas research into wild bee colonies has fallen by the wayside.

The question remains as to why the bee research institutes seem to have devoted themselves exclusively to livestock farming. What could be the underlying motivation be to ignore the extremely important research field of wild colonies and, moreover, to ignore the existing studies and findings?

Another well-known German representative, researcher, and developer of intensive animal husbandry of honey bees takes the public view that the days when honey bees could survive without beekeepers are long gone. He further claims that without the beekeeper, bees would die out in Europe within three years (. . .). In addition, the same person recently published a magazine article in which he defamed a number of people who deal with and know about wild honey bees, including Professor Thomas Seeley and Professor Jürgen Tautz.[21]

How could it be otherwise? This representative, and the bee research institutes conduct their science on bees in conventional livestock husbandry. They have been researching and observing bees in this way for decades. This is comparable to scientific studies of the "natural" behavior of animals in the zoo or in the stable. But how much natural behavior do we get in a zoo or industrial agricultural environment? If we never look beyond the fences of the enclosure and observe the animals in their natural environment, we will never grasp their true natural behavior and will always fail to recognize their potential. Apparently, it is not possible to look or think outside the box (hive). As if there was no Africa outside the zoo, the zoo itself becomes an all-encompassing reality.

THE MOVEMENT NOW COMES FROM WITHIN THE PUBLIC ITSELF

Since the relevant institutes do not seem particularly interested in the topic, numerous monitoring programs based on citizen science have already been set up, which now monitor wild bee colonies on their own. These include Coloss, beetrees.org, beekeeping-revolution.com, freelivingbees.com, Bee Embassy, FreeTheBees, etc. These organizations pursue, among other things, the goal of fighting for species protection for honey bees, which they have so far been illegally denied.

BEEKEEPING REVOLUTION—THE SPECIES PROTECTION PROGRAM FOR HONEY BEES

In 2016, as a research assistant at the University of Würzburg under the direction of Professor Jürgen Tautz, I was commissioned to research the natural living conditions of honey bees in tree cavities, to compare them with living conditions in boxes, and to identify the potential effects on bee health. Remarkably, this pilot research had never been carried out before! Just consider: we actually know more about the natural behavior of the exotic snakes, reptiles, spiders, and other pets that we keep in computer-controlled terrariums in our latitudes than about the most important animal species on earth from an ecological point of view. Scientific research into the viability of honey bees—in which they were completely left to their own devices—revealed that some (after heavy losses) can adapt to local situations and Varroa mites and survive within a few years. Only nature—with the entirety of its selective processes—is able to create viable genetic material that is independent of human influence, and thus able to secure the continued existence of the species in the long term.

In order to be able to give the honey bees back their evolution and thus their future at all, we first had to understand the special physical properties of tree cavities in order to best recreate an equivalent habitat. From there we were able to develop tree-like bee habitats, which enable the bees and other threatened species to live in a species-appropriate way for the first time since the large-scale deforestation of the old trees. These SchifferTrees are offered both commercially and non-commercially.[22] To counter prejudice, I would like to add that I don't earn a single cent from these. The manufacturing companies have received free licenses to produce them.

Species-appropriate beekeeping serves to preserve the species, and does not pursue any economic interests. It is therefore holistically nature-oriented and expressly excludes interventions that have a significant effect on behavior, population development, and biology. Here it is clearly delineated from all other forms of beekeeping, especially honey production.

The movement therefore pursues the goal of restoring the natural balance that has been lost through the clearing of the forests, as a counterweight to human breeding and selection. In the intermediate term, this can only be achieved if the majority of the honey bees' gene pool is returned to natural selection. The entirety of all criteria and facets that make up the survivability of a bee colony in the wild and in its respective region has largely been unexplored. No human-designed breeding program can claim to be as adaptive with regard to the complexity of bee genetics, or be better able to shape the honey bee genome than the processes of natural selection.

REWILDING PROJECTS AND WILD COLONIES

Numerous SchifferTrees were hung in the trees at several locations in spring and summer. Almost all of the empty ones have been populated by natural swarms. These will be scientifically monitored in the coming years. Equally fortunate—bats, starlings, and hornets also moved into some of the cavities. This year alone more than 400 [. . .] such trees were installed in Europe and most of them were populated with bees.

SUMMARY

While clear criteria for animal welfare have already been defined in all other areas of livestock husbandry, so that the various modes of operation can be differentiated, from factory farming to species-appropriate forms of husbandry, this has so far been neglected with bees. On the one hand, the necessary data from tree cavities was missing, on the other hand, bees have no voice and their appearance usually does not tell us what burdens they are carrying, whether external or internal. Modern beekeeping actually no longer has anything to do with the old image of the beekeeper as a friend of the bees and nature, who is rewarded with honey for his diligent care. Addressing these facts also harbors a lot of potential for conflict and is already throwing up a lot of dust.

In addition, the lack of knowledge about the origin of diseases and epidemics ensures that these side effects and problems are incorrectly transferred to all other ways of keeping bees. Even many beekeepers are apparently not aware that the symptoms they spend their efforts combating are caused by the very beekeeping methods they are using.

It is very clear to me how bees are kept today. But I also know that those who have learned or are learning to keep bees had no freedom of choice in their method of operation. But good decisions can only be made on the basis of good information. Only when we understand and comprehend the complexities of the interrelationships can we also have the opportunity to choose alternatives. By glossing over the current situation, we will not be able to bring about any positive changes.

How can it be justified, for example, that we speak of "sustainable environmental education" in schools, while at the same time the student beekeeping program uses medications to subjugate one of the most important species in the world, and is teaching the children (!) how to use them? In this way of keeping bees, not only do the natural needs of the bees fall by the wayside, but also those of the majority of beekeepers themselves! Our own surveys showed that around 70 percent of beekeepers would prefer species-appropriate beekeeping and that honey is not that important to them.

How can it be that the beekeeping education groups claim that honey bees can only be kept in intensive factory farming conditions under medication, while in the same breath, deny that species conservation is viable due to the side effects of those same exploitative conditions?

This development has now led to the fact that a valuable product, historically only available in small quantities, has become a cheap mass-produced product. There are several hundred kilos of honey in every supermarket. The price per glass, which has fallen due to the sheer volume, has led to constant attempts to increase production volumes even further. Some of the conventional representatives of beekeeping are already asking about subsidies and complaining about the low price of their goods. The approximately 500 professional beekeepers in Germany are not the problem—there aren't enough of them to really matter. Rather, the problems are created by the fact that there are around 125,000 so-called recreational beekeepers in addition to the professionals who practice the same intensive factory farming as a "hobby." For the sake of this industry, the evolution of the 45-million-year-old honey bee species was brought to a complete standstill! Insanely enough, honey is valued above the protection and preservation of the bees themselves. This downright fundamentalist perspective has caused most of us to forget that honey bees are actually forest creatures that live in tree hollows. Even governmental bee research institutes, which one would assume serve the bees, participate actively in this policy and, moreover, ignore the state of research on wild colonies.

However, it is not the honey that is of key relevance to the system, but the bees. It is absolutely absurd that there are approximately 570 bee species in Europe under special species protection, and that only one species (which is one of the most important due to its pollination performance) is wrongly denied this protection! We could all easily survive without honey. Without the pollinator insects, however, the most important third of our food will be lost.

This results in the necessary and logical step to reform the training content of the beekeeping schools and adapt it to the existing needs. Not only the craft of keeping bees, but also the resulting effects on the bees themselves must be imparted with equal emphasis. At the same time, it is important to include alternative forms of beekeeping in the core curriculum.

Nobody should fear this reform. We humans do not have to argue, because it is not about blaming each other, but about protecting the species of honey bees and thus preserving them as part of our ecosystem. However, "business as usual" is extremely counterproductive. This includes some institutes and scientists who seriously claim that there is no problem with honey bees, since the beekeeper takes care of them (. . .). To make an analogy, this would be like having no problem with animals that are dying out (e.g., polar bears, rhinos, or tigers), as long as we are still artificially breeding them in the zoo. This limited perspective has little in common with the goals and demands of the "Beekeeping (R) evolution" and numerous other organizations. We are not talking about keeping honey bees alive across the board with the help of manipulative interventions and chemicals in boxes in order to keep them economically viable. We want to return the majority of the *Apis mellifera* gene pool to natural selection and thus restore the lost balance to human breeding and selection. In this way, gene erosion caused by humans can find its way back into balance. The reservoir that is refilled in this way by nature, by which vital genetic material has been created through natural processes since prehistoric times, will ultimately also preserve beekeeping. The approaches do not contradict each other at all, but mutually benefit. All known forms of beekeeping have their justifications and authorizations and require the same level of acceptance.

Compensatory measures according to Section 14 of the Federal Nature Conservation Act could also represent a sensible alternative. It would be conceivable, for example, that one or more species-appropriate hives (e.g., tree cavity simulations) are set up for each managed colony and that swarms that move in there remain untouched. Since we cannot wait many decades or centuries for new natural tree hollows to emerge, we have to compensate for the habitat that has been decimated by deforestation and building with suitable hives. Only natural (species-appropriate) conditions enable natural selection and adaptation. The SchifferTree may provide a corresponding example here. Even in forestry, compensatory measures could lead to an artificial tree cavity being put up for every natural cavity lost in order to stop further shrinking of the habitat.

MONITORING AND RESEARCH OF THE WILD POPULATION

The bee colonies living in the wild in our forests must be recorded and observed in the future. In particular due to the fact that they have been considered extinct for decades, but this has already been invalidated by the current monitoring. These colonies, which survive independently of humans, are of inestimable value for the continued existence of the entire species and require special protection regulations. The concept of numerous projects such as "Beekeeping (R)evolution" aims to further expand the comprehensive recording and observation of these colonies in the coming years with the help of citizen nature lovers who are interested in bees. In this way, monitoring can grow steadily without running into limits imposed by time, manpower, or finances. Such concepts should not only be limited to Germany, but should also be extended to all other countries in which honey bees are considered extinct or threatened in nature.

The data obtained in this way will ultimately provide the template for the reform of the Federal Species Protection Act so that the legal protection status can be extended to wild honey bees. After this step, protected areas or legally defined compensatory measures would also be conceivable.

COURSES IN SPECIES-APPROPRIATE BEEKEEPING

In order to meet the steadily growing demand, we will probably hold regular courses in species-appropriate beekeeping from 2021 onward. In the Hamburg area in particular, I will work to ensure that the conventional beekeeping that is practiced in some general education schools is replaced by a species-appropriate way of keeping bees. You can find detailed information in the Book: *"Evolution der Bienenhaltung – Artenschutz für Honigbienen"*, published by Ulmer Verlag. The coming decades will decide whether we manage to preserve the essential components of the ecosystem in which we live. So, what kind of environment and future do we want to leave our children and grandchildren?

REFERENCES

1. Johann Thür, "Beekeeping: Natural, Simple, and Reliable. Part 1" (1946), translation available at http://www.users.callnetuk.com/~heaf/thur.pdf.
2. Jürgen Tautz, *The Buzz about Bees: Biology of a Superorganism* (Springer: 2008). Data measured from a beehive stand.
3. Arbeitsgemeinschaft der Institute für Bienenforschung E.V., *Stellungnahme zur Konkurrenz zwischen Wildbienen und Honigbienen anlässlich des Positionspapiers des Institutes für Naturkunde aus dem Südwesten* (January 2018) with the title "Wildbienen First" by Ronald Burger, https://deutscherimkerbund.de/userfiles/Wissenschaft_Forschung_Zucht/Stellungnahme_AG_Konkurrenz_Wild-_und_Honigbienen.pdf.
4. Christian Schwägerl, "Was wir über das Insektensterben wissen – und was nicht," *Spectrum.de* (March 2, 2018), www.spektrum.de/wissen/es-gibt-wenig-daten-aber-das-insektensterben-ist-eindeutig-besorgnis-erregend/1548199.
5. Alina Martin and Harry Abraham, "Wildbienen," NABU, www.nabu-krefeld-viersen.de/aktionen-projekte/wildbienen.
6. Torben Schiffer, *Evolution der Bienenhaltung: Artenschutz für Honigbienen* (Ulmer Verlag: 2020).
7. Elmin Taric, Uros Glavinic, Jevrosima Stevanovic, and Branislav Vejnovic, "Occurrence of honey bee (*Apis mellifera L.*) pathogens in commercial and traditional hives," *Journal of Apicultural Research* 58, no. 3 (January 2019): 1–11. DOI: 10.1080/00218839.2018.1554231.
8. Torben Schiffer, *Evolution der Bienenhaltung: Artenschutz für Honigbienen* (Ulmer Verlag: 2020).
9. Johann Thür, "Beekeeping: Natural, Simple, and Reliable. Part 1" (1946), translation available at http://www.users.callnetuk.com/~heaf/thur.pdf.
10. Renata S. Borba, Karen K. Klyczek, Kim L. Mogen, and Marla Spivak, "Seasonal benefits of a natural propolis envelope to honey bee immunity and colony health," *Journal of Experimental Biology* 218 (2015), 3689–99. DOI: 10.1242/jeb.127324.
11. Torben Schiffer, *Evolution der Bienenhaltung: Artenschutz für Honigbienen* (Ulmer Verlag: 2020).

12. Deutscher Imkerbund E. V., *Jahresbericht 2017/2018*, 10–11, https://deutscherimkerbund.de/userfiles/DIB_Pressedienst/Jahresbericht_2017_18.pdf.
13. Felix Hackenbruch, "Honigbienenhaltung hat mit Naturschutz überhaupt nichts zu tun," *Der Tagesspiegel* (July 21, 2019), https://www.tagesspiegel.de/wirtschaft/das-geschaeft-mit-denbienen-honigbienen-haltung-hat-mit-naturschutz-ueberhaupt-nichts-zu-tun/24680722.html.
14. Johannes Wirz, Eva Frey, and Norbert Poeplau, "Varroatoleranz und die Verantwortung der Imker," *Biene Mensch Natur* 28 (Spring/Summer 2020), 12–13, https://www.mellifera.de/blog/biene-mensch-natur-blog/varroatoleranz-und-die-verantwortung-der-imker.html.
15. Deutscher Imkerbund E.V., "Richtlinien für das Zuchtwesen des Deutschen Imkerbundes (ZRL)" (2017), https://deutscherimkerbund.de/userfiles/Wissenschaft_Forschung_Zucht/Zuchtrichtlinien_06_2017_docx.pdf.
16. Goran Mirjanic, Ivana Tlak Gajger, Mica Mladenovic, and Zvonimir Kozaric, "Impact of Different Feed on Intestine Health of Honey Bees," (2013), http://www.resistantbees.com/fotos/estudio/feeding.pdf.
17. R. M. Goodwin, A. Ten Houten, J. H. Perry, "Incidence of American foulbrood infections in feral honey bee colonies in New Zealand," *New Zealand Journal of Zoology* 21 (1994): 285–87. DOI: 10.1080/03014223.1994.9517996.
18. M. E. Miller, "Natural Comb Building", Canadian Bee Journal 43, no. 8 (1935): 216–217.
19. L. Bailey, "Wild Honeybees and Disease," *Bee World* 39 (1958): 92–95.
20. Barbara Locke, "Natural *Varroa* mite-surviving *Apis mellifera* honeybee populations. Apidologie 47 (2016): 467–82. DOI: 10.1007/s13592-015-0412-8.
21. Gerhard Liebig, "Rezensionen Dr. Gerhard Liebig," *Bienenpflege: Die Zeitschrift für den Imker* 9 (2020).
22. www.beenature-project.com

Darwinian Beekeeping: An Evolutionary Approach to Apiculture

Thomas D. Seeley, Cornell University

Evolution by natural selection is a foundational concept for understanding the biology of honey bees, but it has rarely been used to provide insights into the craft of beekeeping. This is unfortunate because solutions to the problems of beekeeping and bee health may come most rapidly if we are as attuned to the biologist Charles R. Darwin as we are to the Reverend Lorenzo L. Langstroth.

Adopting an evolutionary perspective on beekeeping may lead to better understanding about the maladies of our bees, and ultimately improve our beekeeping and the pleasure we get from our bees. An important first step toward developing a Darwinian perspective on beekeeping is to recognize that honey bees have a stunningly long evolutionary history, evident from the fossil record. One of the most beautiful of all insect fossils is that of a worker honey bee, in the species *Apis henshawi,* discovered in 30-million-year-old shales from Germany (Fig. 1). There also exist superb fossils of our modern honey bee species, *Apis mellifera,* in amber-like materials collected in East Africa that are about 1.6 million years old (Engel 1998).

We know, therefore, that honey bee colonies have experienced millions of years being shaped by the relentless operation of natural selection. Natu-

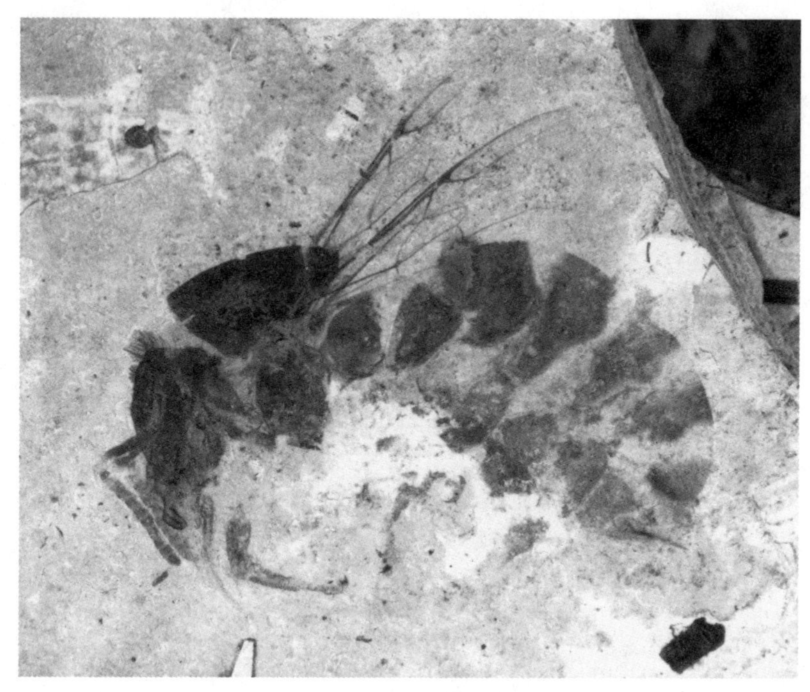

FIG. 1 Photograph of a 30-million-year-old fossil of a worker honey bee in the species *Apis henshawi*. This worker is 0.55 inches long, so its size is close to that of workers in *Apis mellifera*. Photo courtesy of Laurie Burnham.

ral selection maximizes the abilities of living systems (such honey bee colonies) to pass on their genes to future generations. Colonies differ genetically, therefore colonies differ in all the traits that have a genetic basis, including colony defensiveness, vigor in foraging, and resistance to diseases. The colonies best endowed with genes favoring colony survival and reproduction in their locale have the highest success in passing their genes on to subsequent generations, so over time the colonies in a region become well adapted to their environment.

This process of adaptation by natural selection produced the differences in worker bee color, morphology, and behavior that distinguish the 27 subspecies of *Apis mellifera* (e.g., *A.m. mellifera*, A. m. *ligustica*, and A. m. *scutellata*) that live within the species' original range of Europe, western Asia, and Africa (Ruttner 1988). The colonies in each subspecies are precisely adapted to the climate, seasons, flora, predators, and diseases in their region of the world.

Moreover, within the geographical range of each subspecies natural selection produced ecotypes, which are fine-tuned, locally adapted populations. For example, one ecotype of the subspecies *Apis mellifera mellifera* evolved in the Landes region of southwest France, with its biology tightly linked to the massive bloom of heather (*Calluna vulgaris* L.) in August and September. Colonies native to this region have a second strong peak of brood rearing in August that helps them exploit this heather bloom. Experiments have shown that the curious annual brood cycle of colonies in the Landes region is an adaptive, genetically based trait (Louveaux 1973, Strange et al. 2007).

Modern humans (*Homo sapiens*) are a recent evolutionary innovation compared to honey bees. We arose some 150,000 years ago in the African savannahs, where honey bees had already been living for eons. The earliest humans were hunter gatherers who hunted honey bees for their honey, the most delicious of all natural foods. We certainly see an appetite for honey in one hunter-gatherer people still in existence, the Hadza of northern Tanzania. Hadza men spend 4-5 hours per day in bee hunting, and honey is their favorite food (Marlowe et al. 2014).

Bee hunting began to be superseded by beekeeping some 10,000 years ago, when people in several cultures started farming and began domesticating plants and animals. Two regions where this transformation in human history occurred are the alluvial plains of Mesopotamia and the Nile Delta. In both places, ancient hive beekeeping has been documented by archaeologists. Both are within the original distribution of *Apis mellifera*, and both have open habitats where swarms seeking a nest site probably had difficulty finding natural cavities and occupied the clay pots and grass baskets of the early farmers (Crane 1999).

In Egypt's sun temple of King Ne-user-re at Abu Ghorab, there is a stone bas-relief ca. 4400 years old that shows a beekeeper kneeling by a stack of nine cylindrical clay hives (Fig. 2). This is the earliest indication of hive beekeeping and it marks the start of our search for an optimal system of beekeeping. It also marks the start of managed colonies living in circum-

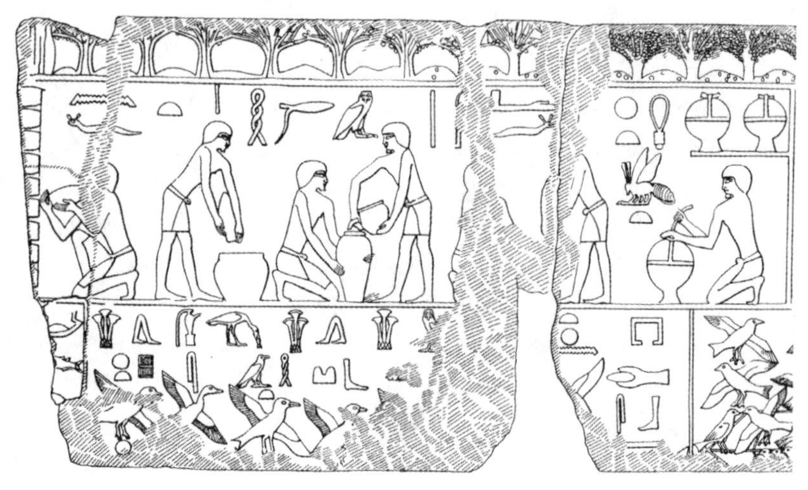

FIG. 2 Earliest known depiction of beekeeping and honey preparation, from the sun temple of King Ne-user-re, at Abu Ghorab, Egypt, built around 2400 BCE. Harvesting honey from a tall stack of cylindrical hives on the left, handling honey in the middle, and packing honey on the right. Drawing based on Fig. 20.3a in Crane (1999).

stances that differ markedly from the environment in which they evolved and to which they were adapted. Notice, for example, how the colonies in the hives depicted in the Egyptian bas-relief lived crowded together rather than spaced widely across the land.

WILD COLONIES VS. MANAGED COLONIES

Today there are considerable differences between the environment of evolutionary adaptation that shaped the biology of wild honey bee colonies and the current circumstances of managed honey bee colonies. Wild and managed live under different conditions because we beekeepers, like all farmers, modify the environments in which our livestock live to boost their productivity. Unfortunately, these changes in the living conditions of agricultural animals often make them more prone to pests and pathogens. In Table 1, I list 20 ways in which the living conditions of honey bees differ between wild and managed colonies, and I am sure you can think of still more.

TABLE 1 **Comparison of the environments in which honey bee colonies lived (and sometimes still do) as wild colonies and those in which they live currently as managed colonies.**

Environment of evolutionary adaptedness	Current circumstances
1. Colonies genetically adapted to location	Colonies not genetically adapted to location
2. Colonies live widely spaced in landscape	Colonies live crowded in apiaries
3. Colonies occupy small (ca 1.5 cu ft) cavities	Colonies occupy large (ca. 3+ cu ft) hives
4. Nest cavity walls have a propolis coating	Hive walls have no propolis coating
5. Nest cavity walls are thick (ca. 4+ in)	Hive walls are thin (ca. 3/4 in)
6. Nest entrance is high & small (ca. 4 sq in)	Nest entrance is low & large (ca. 12 sq in)
7. Nest has 10-25% drone comb	Nest has little (< 5%) drone comb
8. Nest organization is stable	Nest organization is often altered
9. Nest-site relocations are rare	Hive relocations can be frequent
10. Colonies are rarely disturbed	Colonies are frequently disturbed
11. Colonies deal with familiar diseases	Colonies deal with novel diseases
12. Colonies have diverse pollen sources	Colonies have homogeneous pollen sources
13. Colonies have natural diets	Colonies sometimes have artificial diets
14. Colonies are not exposed to novel toxins	Colonies exposed to insecticides & fungicides
15. Colonies are not treated for diseases	Colonies are treated for diseases

continued

Environment of evolutionary adaptedness	Current circumstances
16. Pollen not trapped, honey not taken	Pollen sometimes trapped, honey often taken
17. Beeswax is not removed	Beeswax is removed during honey harvests
18. Bees choose larvae for queen rearing	Beekeepers choose larvae for queen rearing
19. Drones compete fiercely for mating	Queen breeder may select drones for mating
20. Drone brood not removed for mite control	Drone brood sometimes removed and frozen

Difference 1: Colonies are vs. are not genetically adapted to their locations. Each of the subspecies of *Apis mellifera* was adapted to the climate and flora of its geographic range and each ecotype within a subspecies was adapted to a particular environment. Shipping mated queens and moving colonies long distances for migratory beekeeping forces colonies to live where they may be poorly suited. A recent, large-scale experiment conducted in Europe found that colonies with queens of local origin lived longer than colonies with queens of non-local origin (Büchler et al. 2014).

Difference 2: Colonies live widely spaced across the landscape vs. crowded in apiaries. This difference makes beekeeping practical, but it also creates a fundamental change in the ecology of honey bees. Crowded colonies experience greater competition for forage, greater risk of being robbed, and greater problems reproducing (e.g., swarms combining and queens entering wrong hives after mating). Probably the most harmful consequence of crowding colonies, though, is boosting pathogen and parasite transmission between colonies (Seeley & Smith 2015). This facilitation of disease transmission boosts the incidence of disease and it keeps alive the virulent strains of the bees' disease agents.

Difference 3: Colonies live in relatively small nest cavities vs. in large hives. This difference also profoundly changes the ecology of honey bees. Colonies in large hives have the space to store huge honey crops but they also swarm less because they are not as space limited, which weakens natural selection for strong, healthy colonies since fewer colonies reproduce. Colonies kept in large hives also suffer greater problems with brood parasites such as *Varroa* (Loftus et al. 2016).

Difference 4: Colonies live with vs. without a nest envelope of antimicrobial plant resin. Living without a propolis envelope increases the cost of colony defense against pathogens. For example, worker in colonies without a propolis envelope invest more in costly immune system activity (i.e., synthesis of antimicrobial peptides) relative to workers in colonies with a propolis envelope (Borba et al. 2015).

Difference 5: Colonies have thick vs. thin nest cavity walls. This creates a difference in the energetic cost of colony thermoregulation, esp. in cold climates. The rate of heat loss for a wild colony living in a typical tree cavity is 4-7 times lower than for a managed colony living in a standard wooden hive (Mitchell 2016).

Difference 6: Colonies live with high and small vs. low and large entrances. This difference renders managed colonies more vulnerable to robbing and predation (large entrances are harder to guard), and it may lower their winter survival (low entrances get blocked by snow, preventing cleansing flights).

Difference 7: Colonies live with vs. without plentiful drone comb. Inhibiting colonies from rearing drones boosts their honey production (Seeley 2002) and slows reproduction by *Varroa* (Martin 1998), but it also hampers natural selection for colony health by preventing the healthiest colonies from passing on their genes (via drones) the most successfully.

Difference 8: Colonies live with vs. without a stable nest organization. Disruptions of nest organization for beekeeping may hinder colony functioning. In nature, honey bee colonies organize their nests with a precise 3-D organization: compact broodnest surrounded by pollen stores and honey stored above (Montovan et al. 2013). Beekeeping practices that modify the nest organization, such as inserting empty combs to reduce congestion in the broodnest, hamper thermoregulation and may disrupt other aspects of colony functioning such as egg laying by the queen and pollen storage by foragers.

Difference 9: Colonies experience infrequent vs. sometimes frequent relocations. Whenever a colony is moved to a new location, as in migratory beekeeping, the foragers must relearn the landmarks around their hive and must discover new sources of nectar, pollen, and water. One study found that colonies moved overnight to a new location had smaller weight gains in the week following the move relative to control colonies already living in the location (Moeller 1975).

Difference 10: Colonies are rarely vs. frequently disturbed. We do not know how frequently wild colonies experience disturbances (e.g., bear attacks), but it is probably rarer than for managed colonies whose nests are easily cracked open, smoked, and manipulated. In one experiment, Taber (1963) compared the weight gains of colonies that were and were not inspected during a honey flow, and found that colonies that were inspected gained 20-30% less weight (depending on extent of disturbance) than control colonies on the day of the inspections.

Difference 11: Colonies do not vs. do deal with novel diseases. Historically, honey bee colonies dealt only with the parasites and pathogens with whom they had long been in an arms race. Therefore, they had evolved means of surviving with their agents of disease. We humans changed all this when we triggered the global spread of the ectoparasitic mite *Varroa destructor* from eastern Asia, small hive beetle (*Aethina tumida*) from sub-Saharan Africa, and chalkbrood fungus (*Ascosphaera apis*) and acarine mite (*Acarapis woodi*) from Europe. The spread of *Varroa* alone has resulted in the deaths of millions of honey bee colonies (Martin 2012).

Difference 12: Colonies have diverse vs. homogeneous food sources. Some managed colonies are placed in agricultural ecosystems (e.g., huge almond orchards or vast fields of oilseed rape) where they experience low diversity pollen diets and poorer nutrition. The effects of pollen diversity were studied by comparing nurse bees given diets with monofloral pollens or polyfloral pollens. Bees fed the polyfloral pollen lived longer than those fed the monofloral pollens (Di Pasquale et al. 2013).

Difference 13: Colonies have natural diets vs. sometimes being fed artificial diets. Some beekeepers feed their colonies protein supplements ("pollen substitutes") to stimulate colony growth before pollen is available, to fulfill pollination contracts and produce larger honey crops. The best pollen supplements/substitutes do stimulate brood rearing, though not as well as real pollen (http://scientificbeekeeping.com/a-comparative-test-of-the-pollen-sub/) and may result in workers of poorer quality (Scofield and Mattila 2015).

Difference 14: Colonies are not vs. are exposed to novel toxins. The most important new toxins of honey bees are insecticides and fungicides, substances for which the bees have not had time to evolve detoxification mechanisms. Honey bees are now exposed to an ever increasing list of pesticides and fungicides that can synergise to cause harm to bees (Mullin et al. 2010).

Difference 15: Colonies are not vs. are treated for diseases. When we treat our colonies for diseases, we interfere with the host-parasite arms race between *Apis mellifera* and its pathogens and parasites. Specifically, we weaken natural selection for disease resistance. It is no surprise that most managed colonies in North America and Europe possess little resistance to *Varroa* mites, or that there are populations of wild colonies on both continents that have evolved strong resistance to these mites (Locke 2016). Treating colonies with acaracides and antibiotics may also interfere with the microbiomes of a colony's bees (Engel et al. 2016).

Difference 16: Colonies are not vs. are managed as sources of pollen and honey. Colonies managed for honey production are housed in large hives, so they are more productive. However, they are also less apt to reproduce (swarm) so there is less scope for natural selection for healthy colonies. Also, the vast quantity of brood in large-hive colonies renders them vulnerable to population explosions of *Varroa* mites and other disease agents that reproduce in brood (Loftus et al. 2016).

Difference 17: Colonies do not vs. do suffer losses of beeswax. Removing beeswax from a colony imposes a serious energetic burden. The weight-to-weight efficiency of beeswax synthesis from sugar is at best about 0.10 (data of Weiss 1965, analyzed in Hepburn 1986), so every pound of wax taken from a colony costs it some 10 pounds of honey that is not available for other purposes, such as winter survival. The most energetically burdensome way of harvesting honey is removal of entire combs filled with honey (e.g., cut comb honey and crushed comb honey). It is less burdensome to produce extracted honey since this removes just the cappings wax.

Difference 18: Colonies are vs. are not choosing the larvae used for rearing queens. When we graft day-old larvae into artificial queen cups during queen rearing, we prevent the bees from choosing which larvae will develop into queens. One study found that in emergency queen rearing the bees do not choose larvae at random and instead favor those of certain patrilines (Moritz et al. 2005).

Difference 19: Drones are vs. are not allowed to compete fiercely for mating. In bee breeding programs that use artificial insemination, the drones that provide sperm do not have to prove their vigor by competing amongst other drones for mating. This weakens the sexual selection for drones that possess genes for health and strength.

Difference 20: Drone brood is not vs. is removed from colonies for mite control. The practice of removing drone brood from colonies to control *Varroa destructor* partially castrates colonies and so interferes with natural selection for colonies that are healthy enough to invest heavily in drone production.

SUGGESTIONS FOR DARWINIAN BEEKEEPING

Beekeeping looks different from an evolutionary perspective. We see that colonies of honey bees lived independently from humans for millions of years, and during this time they were shaped by natural selection to be skilled at surviving and reproducing wherever they lived, in Europe, western Asia, or Africa. We also see that ever since humans started keeping bees in hives, we have been disrupting the exquisite fit that once existed between honey bee colonies and their environments. We've done this in two ways: 1) by moving colonies to geographical locations to which they are not well adapted, and 2) by managing colonies in ways that interfere with their lives but that provide us with honey, beeswax, propolis, pollen, royal jelly, and pollination services.

What can we do, as beekeepers, to help honey bee colonies live with a better fit to their environment, and thereby live with less stress and better health? The answer to this question depends greatly on how many colonies you manage, and what you want from your bees. A beekeeper who has a few colonies and low expectations for honey crops, for example, is in a vastly different situation than a beekeeper who has thousands of colonies and is earning a living through beekeeping.

For those interested, I offer 10 suggestions for bee-friendly beekeeping. Some have general application while others are feasible only for the backyard beekeeper.

1. *Work with bees that are adapted to your location.* For example, if you live in New England, buy queens and nucs produced up north rather than queens and packages shipped up from the south. Or, if you live in a location where there are few beekeepers, use bait hives to capture swarms from the wild colonies living in your area. (Incidentally, these swarms will build you beautiful new combs, and this will enable you to retire old combs that could have heavy loads of pesticide residues and pathogen spores/cells.) The key thing is to acquire queens of a stock that is adapted to your climate.
2. *Space your hives as widely as possible.* Where I live, in central New York State, there are vast forests filled with wild honey bee colonies spaced

roughly a half mile apart. This is perhaps ideal for wild colonies but problematic for the beekeeper. Still, spacing colonies just 30-50 yards apart in an apiary greatly reduces drifting and thus the spread of disease.

3. *House your bees in small hives.* Consider using just one deep hive body for a broodnest and one medium-depth super over a queen excluder for honey. You won't harvest as much honey, but you will likely have reduced disease and pest problems, particularly *Varroa*. And yes, your colonies will swarm, but swarming is natural and research shows that it promotes colony health by helping keep *Varroa* mite populations at safe levels (see Loftus et al 2016).
4. *Roughen the inner walls of your hives, or build them of rough-sawn lumber.* This will stimulate your colonies to coat the interior surfaces of their hives with propolis, thereby creating antimicrobial envelopes around their nests.
5. *Use hives whose walls provide good insulation.* These might be hives built of thick lumber, or they might be hives made of plastic foam. We urgently need research on how much insulation is best for colonies in different climates, and how it is best provided.
6. *Position hives high off the ground.* This is not always doable, but if you have a porch or deck where you can position some hives, then perhaps it is feasible. We urgently need research on how much entrance height is best in different climates.
7. *Let 10-20% of the comb in your hives be drone comb.* Giving your colonies the opportunity to rear drones can help improve the genetics in your area. Drones are costly, so it is only the strongest and healthiest colonies that can afford to produce legions of drones. Unfortunately, drone brood also fosters rapid growth of a colony's population of *Varroa* mites, so providing plentiful drone comb requires careful monitoring of the *Varroa* levels in your hives (see suggestion 10, below).
8. *Minimize disturbances of nest organization.* When working a colony, replace each frame in its original position and orientation. Also, avoid inserting empty frames in the broodnest to inhibit swarming.
9. *Minimize relocations of hives.* Move colonies as rarely as possible. If you must do so, then do so when there is little forage available.

10. *Refrain from treating colonies for Varroa.* WARNING: This last suggestion should only be adopted if you can do so carefully, as part of a program of extremely diligent beekeeping. If you pursue treatment-free beekeeping without close attention to your colonies, then you will create a situation in your apiary in which natural selection is favoring virulent *Varroa* mites, not *Varroa*-resistant bees. To help natural selection favor *Varroa*-resistant bees, you will need to monitor closely the mite levels in all your colonies and euthanize those whose mite populations are skyrocketing long before these colonies collapse. By preemptively killing your *Varroa*-susceptible colonies, you will accomplish two important things: 1) you will eliminate your colonies that lack *Varroa* resistance and 2) you will prevent the "mite bomb" phenomenon of mites spreading en masse to other colonies. If you don't perform these preemptive killings, then even your most resistant colonies, living near the collapsing one(s) could become overrun with mites and die. If this happens, then there will be no natural selection for mite resistance in your apiary. Failure to perform preemptive killings can also spread virulent mites to your neighbors' colonies and even to the wild colonies in your area that are slowly evolving resistance on their own. If you are not willing to euthanize your mite-susceptible colonies, then you will need to treat them to kill the mites and then requeen them with a queen of mite-resistant stock.

TWO HOPES

I hope you have found it useful to think about beekeeping from an evolutionary perspective. If you are interested in pursuing beekeeping in a way that is centered less on treating a bee colony as a honey factory, and more on nurturing the lives of honey bees, then I encourage you to consider what I call Darwinian Beekeeping. Others call it Natural Beekeeping, Apicentric Beekeeping, and Bee-friendly Beekeeping (Phipps 2016). Whatever the name, its practitioners view a honey bee colony as a complex bundle of adaptations shaped by natural selection to maximize a colony's survival and reproduction in competition with other colonies and other organisms (predators, parasites, and pathogens). It seeks to foster colony health by letting the bees live as naturally as possible, so they can make full use of the toolkit of adaptations that they have acquired over the last 30 million years. Much remains to be learned about this toolkit—How exactly do colonies benefit from better nest insulation? Do colonies tightly seal their nests with propolis in autumn to have an in-hive water supply (condensate) over winter? How exactly do colonies benefit from having a high nest entrance? The methods of Darwinian Beekeeping are still being developed, but fortunately, apicultural research is starting to embrace a Darwinian perspective (Neumann and Blacquiere 2016).

I hope too that you will consider giving Darwinian Beekeeping a try, for you might find it more enjoyable than conventional beekeeping, especially if you are a small-scale beekeeper. Everything is done with bee-friendly intentions and in ways that harmonize with the natural history of *Apis mellifera*. As someone who has devoted his scientific career to investigating the marvelous inner workings of honey bee colonies, it saddens me to see how profoundly—and ever increasingly—conventional beekeeping disrupts and endangers the lives of colonies. Darwinian Beekeeping, which integrates respecting the bees and using them for practical purposes, seems to me like a good way to be responsible keepers of these small creatures, our greatest friends among the insects.

ACKNOWLEDGMENTS

I thank Mark Winston and David Peck for many valuable suggestions that improved early drafts of this article. Attending the Bee Audacious Conference in December 2016 is what inspired my thinking on Darwinian Beekeeping, so I also thank Bonnie Morse and everyone else who made this remarkable conference a reality.

REFERENCES

Borba, R.S., K.K. Klyczek, K.L. Mogen and M. Spivak. 2015. Seasonal benefits of a natural propolis envelope to honey bee immunity and colony health. Journal of Experimental Biology 218: 3689-3699.

Büchler, R, C. Costa, F. Hatjina and 16 other authors. 2014. The influence of genetic origin and its interaction with environmental effects on the survival of *Apis mellifera* L. colonies in Europe. Journal of Apicultural Research 53:205-214.

Crane, E. 1999. *The world history of beekeeping and honey hunting.* Duckworth, London.

Di Pasquale, G., M. Salignon, Y. LeConte and 6 other authors. 2013. Influence of pollen nutrition on honey bee health: do pollen quality and diversity matter? PLoS ONE 8(8): e72106.

Engel, M.S. 1998. Fossil honey bees and evolution in the genus *Apis* (Hymenoptera: Apidae). Apidologie 29:265-281.

Engel, P, W.K. Kwong, Q. McFrederick and 30 other authors. 2016. The bee microbiome: impact on bee health and model for evolution and ecology of host-microbe interactions. mBio 7(2): e02164-15.

Hepburn, H.R. 1986. *Honeybees and wax.* Springer-Verlag, Berlin.

Locke, B. 2016. Natural *Varroa* mite-surviving *Apis mellifera* honeybee populations. Apidologie 47:467-482.

Loftus, C.L., M.L. Smith and T.D. Seeley. 2016. How honey bee colonies survive in the wild: testing the importance of small nests and frequent swarming. PLoS ONE 11(3): e0150362.

Louveaux, J. 1973. The acclimatization of bees to a heather region. Bee World 54:105-111.

Marlowe, F.W., J.C. Berbesque, B. Wood, A. Crittenden, C. Porter and A. Mabulla. 2014. Honey, Hadza, hunter-gatherers, and human evolution. Journal of Human Evolution 71:119-128.

Martin, S.J. 1998. A population model for the ectoparasitic mite *Varroa jacobsoni* in honey bee (*Apis mellifera*) colonies. Ecological Modelling 109:267-281.

Martin, S.J., A.C. Highfield, L. Brettell and four other authors. 2012. Global honey bee viral landscape altered by a parasitic mite. Science 336: 1304-1306

Mitchell, D. 2016. Ratios of colony mass to thermal conductance of tree and man-made nest enclosures of *Apis mellifera*: implications for survival, clustering, humidity regulation and *Varroa destructor.* International Journal of Biometeorology 60:629-638.

Moeller, F.E. 1975. Effect of moving honeybee colonies on their subsequent production and consumption of honey. Journal of Apicultural Research 14:127-130.

Montovan, K.J., N. Karst, L.E. Jones and T.D. Seeley. 2013. Local behavioral rules sustain the cell allocation pattern in the combs of honey bee colonies (*Apis mellifera*). Journal of Theoretical Biology 336:75-86.

Moritz, R.F.A., H.M.G. Lattorff, P. Neumann and 3 other authors. 2005. Rare royal families in honey bees, *Apis mellifera*. Naturwissenschaften 92:488-491.

Mullin, C.A., M. Frazier, J.L. Frazier and 4 other authors. 2010. High levels of miticides and agrochemicals in North American apiaries: implications for honey bee health. PLoS ONE 5(3): e9754.

Neumann, P. and T. Blacquiere. 2016. The Darwin cure for apiculture? Natural selection and managed honeybee health. Evolutionary Applications 2016: 1-5. DOI:10.1111/eva.12448

Phipps, J. 2016. Editorial. Natural Bee Husbandry 1:3.

Ruttner, F. 1988. *Biogeography and Taxonomy of Honeybees*. Springer Verlag, Berlin.

Scofield H.N., Mattila H.R. 2015. Honey bee workers that are pollen stressed as larvae become poor foragers and waggle dancers as adults. PLoS ONE 10(4): e0121731.

Seeley, T.D. 2002. The effect of drone comb on a honey bee colony's production of honey. Apidologie 33:75-86.

Seeley, T.D. and M.L. Smith. 2015. Crowding honeybee colonies in apiaries can increase their vulnerability to the deadly ectoparasite *Varroa destructor*. Apidologie 46:716-727.

Strange, J.P., L. Garnery and W.S. Sheppard. 2007. Persistence of the Landes ecotype of *Apis mellifera mellifera* in southwest France: confirmation of a locally adaptive annual brood cycle trait. Apidologie 38:259-267.

Taber, S. 1963. The effect of disturbance on the social behavior of the honey bee colony. American Bee Journal 103 (Aug):286-288.

Weiss, K. 1965. Über den Zuckerverbrauch und die Beanspruchung der Bienen bei der Wachserzeugung. Zeitschrift für Bienenforschung 8:106-124.

Index

* *Italics* are used to indicate illustrations; **Bold** is used to indicate tables.

Index